Advanced Fixture Design for FMS

Advanced Manufacturing Series
Series Editor:D.T. Pham

Other titles in the series:

A Systems Approach to AMT Deployment
Eds. D.R. Towill and J. Cherrington

Human-Intelligence-Based Manufacturing
Ed. Y. Ito

Intelligent Manufacturing: Programming Environments for CIM
Eds. W.A. Gruver and J.C. Boudreaux

Automatic Supervision in Manufacturing
Ed. M. Szafarczyk

Modern Manufacturing: Information Control and Technology
Eds. M. B. Zaremba and B. Prasad

ADVANCED FIXTURE DESIGN FOR FMS

A.Y.C. Nee, K. Whybrew
and A. Senthil kumar

With 105 Figures

Springer-Verlag
London Berlin Heidelberg New York
Paris Tokyo Hong Kong
Barcelona Budapest

A.Y.C. Nee
Department of Mechanical and Production Engineering
National University of Singapore
10 Kent Ridge Crescent, Singapore 0511

K. Whybrew
Department of Mechanical Engineering
University of Canterbury
Private Bag 4800
Christchurch, New Zealand

A. Senthil kumar
Department of Mechanical and Production Engineering
National University of Singapore
10 Kent Ridge Crescent, Singapore 0511

Series Editor
D.T. Pham
University of Wales College of Cardiff
School of Electrical, Electronic and Systems Engineering
P.O. Box 917, Cardiff CF2 1XH, UK

ISBN-13:978-1-4471-2119-0 e-ISBN-13:978-1-4471-2117-6
DOI: 10.1007/978-1-4471-2117-6

British Library Cataloguing in Publication Data
A catalogue record for this book is available from the British Library

Library of Congress Cataloging-in-Publication Data
A catalog record for this book is available from the Library of Congress

Softcover reprint of the hardcover 1st edition 1995

Typesetting: Camera ready by authors

69/3830-543210 Printed on acid-free paper

AYCN — To my wife Siew Kheng and daughters Joanne, Julianne, Suzanne for their patience and understanding

KW — To Linda for all her help

ASK — To my parents Prof Anantharajan and Ms Vijayalakshmi for their help and concern

Preface

During a visit that KW made to the university of Wales at Cardiff in 1991, Professor Pham Duc Truong suggested that a book on Advanced Fixture Design Methodologies should be published under the Advances in Manufacturing series. Later that year KW and his two co-authors, AYCN and ASK, had the pleasure of working together at the National University in Singapore.

We soon realised that although we have the same research interests, that is automation of fixture design and process planning, we had different but complementary perspectives on the problem. At NUS the emphasis has been application of expert systems and feature recognition to synthesise fixture designs. AYCN and ASK have developed an expert system linked to a solid modeller to design fixtures based on machining technology and optimising the number of set-ups to manufacture a workpiece. At the University of Canterbury KW has concentrated on design of assemblies of modular fixture elements and solving problems of connectivity and interference and also sequencing of operations to avoid tolerance problems. Since AYCN and ASK were also planning a book on computer-aided fixture design, there were obvious benefits if we combined our knowledge and resources.

We intend this book to be of interest to practising engineers and researchers involved in process planning and fixture design. As the title suggests the emphasis is flexible manufacturing but the topics covered are not restricted to FMS. In the process of developing this book we have taken a fresh look at the fixture design process and this we hope will be of value to beginners in the art and science of fixture design as well as developers of automated systems.

This book is largely a distillation of the research of a number of our students. We are pleased to acknowledge their contribution, in particular the work of Bryon Ngoi Kok Ann and Yongyooth Sermsuti-Anuwat at University of Canterbury and Tung Kuo Hua, Chan Tee Juay and Puah Kok Yong at National University of Singapore.

We have also drawn heavily on the literature and we gratefully acknowledge the contribution of the many researchers who have added to our understanding of the problems of fixture design. We are particularly indebted to those who have allowed us to reproduce parts of their work. We would also like to thank Imke Mowbray at Springer-Verlag for her help in coordinating the production of this book.

A.Y.C. Nee
National University of Singapore

Ken Whybrew
University of Canterbury

A. Senthil kumar
National University of Singapore

Contents

1 Fixture and Flexible Manufacturing Systems

1.1 Introduction

Fixtures are an essential part of manufacturing production. They are needed whenever a component must be located and held with respect to a machine-tool or measuring device, or with respect to another component, as for instance in assembly or welding. This study concentrates on fixtures for machining and, in particular, fixtures for use in flexible manufacturing systems (FMS).

In machining, the function of a fixture is to ensure that the component being manufactured is held in such a way that the design specification can be satisfied. A fixture does this by:

- providing accurate and repeatable location of the datum surfaces of the workpiece with respect to the axes of the machine-tool;
- resisting motion, deflection and distortion of the workpiece under the action of the cutting tool.

Jigs are similar to fixtures, and indeed fixtures are often referred to by the generic term *jigs and fixtures*, but jigs provide an additional function:

- location of the cutting tool with respect to the workpiece.

With numerically controlled machines the use of jigs has become almost redundant and they will not be considered here.

Many excellent books have been published on design of fixtures. Most of them are aimed at the practitioner - tool designers and tool makers - and provide exemplars for particular applications without establishing a rigorous theoretical basis for design. Notable exceptions are the works by Eary and Johnson [1] and Hoffman [2], in which the requirements of workpiece control are established as the primary objectives for fixture design.

Pressures on the manufacturing industry during the 1980s have led to the development of many new techniques which come under the general description *advanced manufacturing technology* (AMT). Fixtures play a crucial role in these new technologies and they have been the subject of intensive research. This book is an attempt to bring the results of this research together and provide a rational approach to selection and design of fixtures.

1.2 AMT and Fixtures

Manufacturing industry is under two pressures:

- to maximise rate of return on investment;
- to reduce leadtime for introducing new products.

Fixture design has an important bearing on the techniques to achieve these goals.

1.2.1 Fixture Design and Rate of Return on Investment

A simple way of assessing the efficiency of an investment is to consider the rate of return on the invested capital. Rate of return is defined by de Garmo [3] as:

$$\text{rate of return} = \frac{\text{annual net profit}}{\text{invested capital}}$$

$$\text{rate of return} = \frac{[\text{selling price-production cost}] \times \text{no.produced}}{\text{fixed and liquid assets} + \text{stock-inventory} + \text{work-in-progress}}$$

Advanced manufacturing techniques such as group technology (GT), flexible manufacturing systems (FMS), and just-in-time production (JIT) have been introduced to maximise the rate of return. Their main strategy is to decrease invested capital through decrease in work in progress and stock inventory. A detailed study of these techniques can be found in Groover [4]. They are summarised here in relation to fixture design.

In FMS, work-in-progress is decreased by reducing the handling time and the waiting time between operations. Handling time is reduced by using automatic transport and handling equipment. Waiting time between operations is reduced by facilitating rapid change of workpiece type. FMS therefore allow flexibility in workpiece scheduling. For full flexibility in scheduling the FMS must provide:

- mix flexibility - the ability of the system to accommodate variety of components and materials;
- route flexibility - the ability to swap production to different machines;
- volume flexibility - the ability of the system to cope with changes in the required output, with the ideal system being able to economically produce single components.

This is referred to as *short term flexibility*. Mix flexibility, route flexibility and volume flexibility are all dependent on fixture design. Automatic handling and flexible scheduling have the additional benefit of increased machine utilisation and hence lower capital investment in plant for a given output.

Most FMS work with lower manning levels than conventional manufacturing systems, and many will work at least one shift completely unmanned. Production costs are lower bringing additional increase to rate of return. Unmanned automation places further demands on fixture design. The type of fixture used in FMS is dependent on the automation strategy, and the fixture design must be considered as an integral part of material-handling and transport. Williams [5] identifies two main types of FMS:

- cell based
- monolithic

Cell based FMS consist of a number of locally autonomous cells with one or more CNC machines and an intra-cell materials-handling system using either a robot or a pallet carousel and pallet exchanger. Work pieces are delivered to and collected from the cell by an inter-cell material transport system, usually an automatic guided vehicle system (AGV), common to the whole FMS. Material is delivered to the cell loosely located on wooden pallets. Intra-cell handling may take two forms. One approach is for a cell operator to manually establish fixtures on a pallet carousel and manually load components to be machined into the fixtures. A pallet exchanger then automatically transfers the pallet to the machine tool according to the manufacturing schedule. In this way one operator working one shift may service several cells which may work one or two shifts unattended. An example of a typical cell is shown in Figure 1.1.

This approach gives high flexibility but very high fixture numbers as work in progress (WIP) is held in the fixtures. If more than one component is required duplicate fixtures are needed. Large carousels can only be avoided if the machining time is comparatively long. If the skill of the operator can be relied upon, fixtures of simpler design to those demanded for fully automatic operation can be used. This method of intra-cell workpiece handling is often used to produce medium sized prismatic components on milling machines.

The alternative approach is to use robot workpiece loading. Components are either delivered oriented and in a pre-determined pattern, or they are manually positioned on a workpiece carousel. The robot loader picks up and transfers the workpieces to the machine tool and positions them in a fixture already established on the machine tool. Automatic clamps are used to finally position the workpiece in the fixture. An example of this type of cell is shown in Figure 1.2.

Using this approach a higher degree of automation is achieved and components with relatively short machining times can be processed. Multi-component batches of parts can be machined without duplicate fixtures, but unless some form of automatic fixture exchange is used or the fixtures can accommodate

Figure 1.1 Milling cell with a pallet carousel (Courtesy of OKUMA Corporation)

Figure 1.2 Lathe cell with robot loading of workpieces

a range of components the FMS will have very limited flexibility. Severe demands are placed on fixture design to ensure correct automatic placement of the

component and some form of closed loop surveillance of the fixture is also usual. This kind of intra-cell handling is commonly used in turning cells and with small milling machines.

In *monolithic FMS* loading and unloading are usually done off-line at a central fixture service area. Precision pallets and fixtures are retrieved from an automatic retrieval and storage warehouse and the fixture established on the pallet. When scheduled, a part blank is mounted in the fixture by the operator and the pallet/fixture/part combination directed to a delivery station to await collection by an AGV. The AGV delivers the loaded pallet to the appropriate machine tool where a pallet exchanger transfers it to the machine. The exchanged pallet with the previously machined component is then returned to the fixture service area where the component is removed and delivered to the warehouse. The fixture is then stripped from the pallet, cleaned and returned to storage, or if the operation is to be repeated the pallet/fixture combination is cleaned to remove remaining swarf and chips and the next component blank mounted in the fixture. A typical fixture servicing area is shown in Figure 1.3.

This approach to FMS gives maximum mix flexibility and also, provided

Figure 1.3 Monolithic FMS fixture service area (Courtesy of Mori Seiki Ltd.)

the pallets are common to all machines, maximum route flexibility. A disadvantage with this kind of FMS is longer part handling time compared with the cell type of FMS. This can limit volume flexibility when dealing with small components with short machining times. To maintain machine utilization the machining time must be longer than the total handling and transport time. With small components the machining time per pallet can be increased by using multi-workpiece fixtures, but this will obviously reduce the volume flexibility. An example of a multi-workpiece fixture is shown in Figure 1.4. The workpieces may be duplicates of the same component but are often different components that are used together in a particular product.

Figure 1.4 Multi-workpiece milling fixture

With current technology all of the assembly work in fixture service is manual and hence there is some trade-off of operator skill for simple fixtures, but it is the intention of many FMS builders to fully automate all of the fixture service activity. This will have important consequences for fixture design.

JIT is a manufacturing philosophy which ensures that WIP and stock inventory are kept to a minimum by requiring that a component is manufactured only just before it is required for a subsequent process. FMS is a common means to achieve the necessary response from the manufacturing facility. A consequence of JIT is that components must also be manufactured 'right first time' otherwise there is a danger that JIT becomes JTL (just too late)! Right first time production requires a high level of quality assurance. Such assurance is only possible with carefully designed process plans and fixtures which ensure repeatable location of the datum surfaces identified in such plans. Nixon [6, p162], has reported that over 40% of rejects can be directly attributed to poorly designed fixtures.

1.2.2 Fixture Design and Production Leadtime

A characteristic of most FMS built in the last decade, and in particular the very successful Japanese FMS has been the heavy dependence on special purpose fixtures to achieve the requirements of scheduling flexibility and quality. Fixtures are considered to be an integral part of the FMS and represent a major proportion of the total investment, in some cases over 20%. In general each set-up for every component has its own fixture. Scheduling flexibility is achieved by providing the means to quickly retrieve a fixture and establish it on the FMS. FMS provide short term flexibility, but in general their long term flexibility (or *reconfigurability*) is poor. Conversations with managers of some of the best known Japanese FMS indicate that it may take as long as three months to introduce a new component into the FMS, and several European FMS are considered economic failures because of poor long term flexibility. This is in conflict with modern manufacturing's second objective of bringing new products to market in the shortest possible time. Most of the developments described in this book are intended to increase the reconfigurability of fixture set-ups.

Concurrent engineering and computer integrated manufacturing (CIM) are becoming accepted as essential management techniques for minimising leadtime. Computer-aided process planning (CAPP) used within the framework of concurrent engineering is the interface between design and manufacture. Fixture design is concerned with selection of surfaces on the workpiece for locating and clamping during machining and must therefore be seen as an essential part of process planning. CIM is not possible without CAPP and no CAPP system is complete without a computer aided fixture design module. Recent developments in CAD of fixtures are reviewed in Chapter 5.

1.3 Fixture Strategies For FMS

The ideal fixture for FMS should provide the essential functions of location and support and be capable of being reconfigured to accommodate every workpiece processed by the FMS. This is not possible with current technology.

The fixtures used in conventional manufacturing can be considered as falling into three categories:

- commercial standard workholding devices
- special purpose fixtures
- factory standard fixtures

Commercial standard workholding devices include machine vices, chucks, faceplates, angle-blocks, vee-blocks, clamps, etc. Using combinations of these simple devices, a skilled tradesman can construct a fixture for most workpieces. This fixturing strategy is common in low volume jobbing production, toolmaking, and in maintenance workshops. It has the advantage of using highly versatile equipment and a low capital investment in fixtures. The leadtime to introduce a new component is short in comparison with special purpose fixtures, but there are many disadvantages which prevent this fixture strategy being used for FMS. Changing from one fixture to another is slow because the fixture must be rebuilt. This type of fixture is heavily reliant on operator skill to accurately load the workpiece and for alignment of the fixture with the machine-tool axes, hence they are not suited to automation.

Special purpose fixtures are specially designed and manufactured to hold and locate a specific component for a specific operation. They are completely non-versatile and generally cannot be used for other components or operations. They are used in large volume production where the capital investment can be spread over large production numbers, and high levels of automation of clamping and workpiece location are economically viable. Such fixtures are specially designed to exactly match the requirements of the process plan and ensure that repeatable high accuracy is achieved. Because they can be fully automated for unmanned operation they are also a common fixturing strategy for FMS. The drawback of special purpose fixtures is the high capital cost of the fixtures themselves and the cost of storing the enormous number of fixtures required for each and every operation. In effect this means that FMS can only be used for repeated batch production where the total number of components is sufficient to offset the cost of fixtures. Simple special purpose fixtures are also used for conventional batch manufacture, but they are a major cost component.

Factory standard fixtures are a fixture strategy used in conjunction with group technology as a response to the problems of batch manufacture. Special purpose fixtures are designed to accommodate a range of components from a family instead of just a single workpiece. These fixtures are designated as 'factory standard fixtures'. Group technology techniques are applied during the design stage of a new product to ensure that no unnecessary variety is introduced and to enforce concurrence with family standard design features. In this way the total number of fixtures is reduced, and also the time spent in changing components in the same family is dramatically reduced. Factory standard fixtures and variety reduction can give considerable benefits to rate of return and clearly have an important role in FMS.

These conventional strategies do not adequately satisfy the fixture requirements of FMS. Special purpose fixtures used in conjunction with off-line set-up and pallet exchangers give satisfactory short term flexibility, automation and accuracy but only at great cost. The inherent lack of versatility of special purpose fixtures frustrates the second FMS ideal of long term flexibility and ability to quickly introduce new products. So long as they are dependent on special purpose fixtures, FMS will never be an economical manufacturing option for jobbing or short run products.

Factory standard fixtures and group technology impose restraints on designers which may be considered to be unacceptable. This strategy will only work for a captive manufacturing facility servicing a particular product range. It is impossible for an FMS employing this strategy to accept components that do not fit existing part families.

Flexible Fixturing is a generic term for several new fixture techniques which combine the long term flexibility of commercial standard workholding devices with the advantages of special purpose fixtures. Chapter 4 provides an overview of Flexible Fixture methodologies. The main categories are:

- Modular fixture kits
- Phase-change fixtures
- Programmable fixtures

Modular fixture kits are a methodology that is an extension of fixture building using standard workholding devices. Modular fixture kits are made up of a collection of standard clamps, locator components, and modular blocks from which a fixture body to support the clamps and locators can be constructed. These standard components are assembled on a baseplate with either 'T' slots or a grid of holes to create a 'special' fixture. When finished with, the fixture can be dismantled and the parts returned to stock.

Phase change fixtures generally form a supporting structure for the component being machined from an enveloping matrix of phase changing material. The techniques have a striking similarity to the wax dolly used for centuries by jewellers to hold the pieces on which they are working. In the modern version the component is located by some external means and then enveloped in the fluid matrix (eg a low melting point metal) in a mould or container. The matrix is then solidified to form a rigid supporting structure which has standardised location surfaces to fit a standard fixture. The component is removed by re-fluidising the matrix which can then be recycled. Phase change fixtures are commonly used for components that do not have well defined location planes, such as turbine blades.

Programmable fixtures are able to be automatically re-configured to accept different components. Clamps and locators are moved and if necessary changed under programmed control.

By using flexible fixtures and automating the fixture design process using CAD, many of the problems of poor reconfigurability inherent in the present generation of FMS can be overcome without sacrificing essential short-term flexibility. This book will present a summary of these techniques.

1.4 References

1 Eary DF, Johnson GE. Process engineering for manufacturing. Prentice-Hall, Englewood Cliffs, NJ, 1962

2 Hoffman GE (ed). Fundamentals of tool design, 2nd edn. Society of Manufacturing Engineers, Dearborn, Mich, 1984

3 de Garmo EP. Engineering economics, 4th edn. Macmillan, New York,1967

4 Groover MP. Automation production systems and computer integrated manufacturing. Prentice- Hall, Englewood Cliffs, NJ, 1987

5 Williams DJ. Manufacturing systems an introduction to the technologies. Open University Press, 1988

6 Nixon F. Managing to achieve quality and reliability. McGraw-Hill, Maidenhead, 1971

2 Fixture Design Fundamentals

2.1 Introduction

Before new types of fixtures or methods of automating design of fixtures can be considered, it is necessary to establish a fundamental design rationale.

Until comparatively recently fixture design could best be described as an art which owed very little to mechanical engineering science. Fixture designers usually had a background as skilled toolmakers and brought a wealth of practical experience to fixture design. This practical approach to design is reflected in many of the text books on fixture design listed in the Bibliography. They provide a valuable resource of the collected knowledge and experience of many first class tool designers and it is not the intention of this book to duplicate this work, but rather to show how this knowledge can be effectively applied within a framework of advanced design techniques.

Pressures on contemporary industry are making the traditional methods of fixture design obsolete. Designs based on past experience tend to be conservative. They may work effectively but are unlikely to be optimised to satisfy the demands of modern manufacturing. Shortage of skilled labour, demands for increased design productivity and shorter design lead-time conspire to make traditional methods unacceptable. Automation of the design process using computer-aided design provides a solution.

Computer-aided design systems are commonly used as general design aids. The designer provides the creative input to synthesise a new product and critical analysis of the design as it progresses. All knowledge concerning the design environment and procedures must come from the designer and the design is hence limited by the knowledge of the designer. The CAD system serves the designer only as a tool to create and manipulate geometric models of the emerging design, and for communicating the design to other users of the design information. A CAD system can be used in this manner to design fixtures but many potential benefits of CAD will not be realised.

If the class of product, the design procedures and the criteria for decision making are well enough understood, a computer can also be used to store and manipulate knowledge of the design environment [1]. This knowledge used in conjunction with the CAD system may further aid designers by providing a logical framework for the design, guiding them through the design process, providing generic forms of the design and constructive critical analysis of the design as it progresses. The designer may also be constrained by the system to conform to factory standards. The ideal, to which much current research is directed, is a CAD system which will automate the entire process, with the designer interacting with the system to provide information and make decisions where design knowledge is incomplete. Chapter 5 will describe various systems that bring fixture design closer to this goal.

Computer-aided design is impossible if the design process is not thoroughly understood. This chapter will provide a rational basis for fixture design and in particular CAD of fixtures.

2.2 Definitions

The following terms will be used throughout this book:

- *Locator*

A locator is usually a fixed part of a fixture, the purpose of which is to restrict movement of the workpiece being fixtured. A theoretical locator, represented by a small triangle symbol, will prevent movement of the workpiece in one direction of one degree of freedom. This is illustrated in Figure 2.1, where the locator is preventing motion in the x direction.

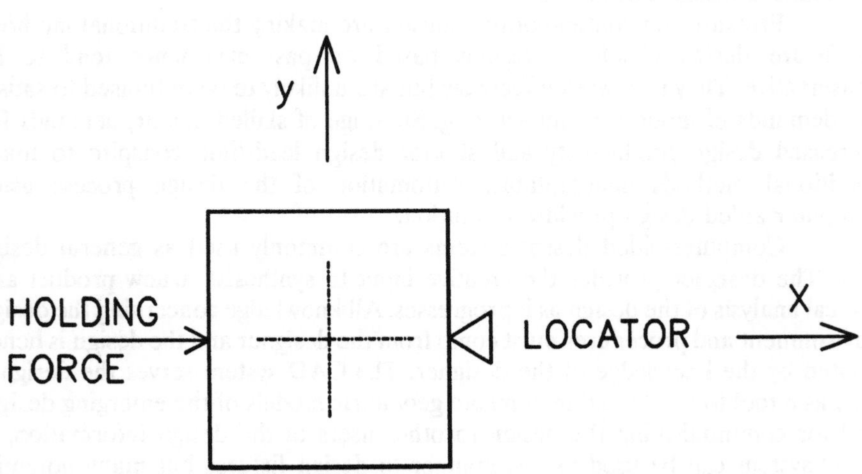

Figure 2.1 Translation location

A locator not aligned with an instantaneous centre of rotation will prevent rotation about that centre. This is illustrated in Figure 2.2.

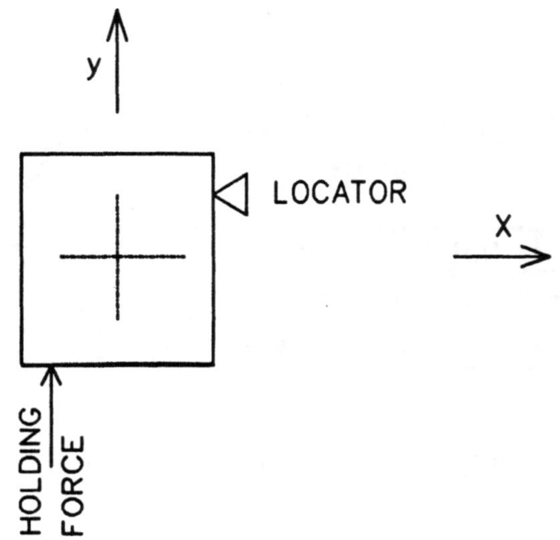

Figure 2.2 Rotation location

For first operations when the surface being located is pre-machined and for second and subsequent operations the locator is invariably fixed. When the location surface for the first operation is not machined and is not well defined, it is sometimes necessary to use adjustable locators. Some practical examples of locators are shown in Figure 2.3.

- *Clamp*

A clamp is a moveable part of a fixture, the purpose of which is to provide a holding force. The holding force may hold the workpiece being fixtured against a locator by preventing motion in the opposite direction or provide a moment preventing rotation about some instantaneous centres. A clamp is represented by an arrow in the direction of the line of action of the holding force it is intended to provide. This is illustrated in Figures 2.1 & 2.2. Some practical examples of clamps are shown in Figure 2.4.

- *Support*

A support is a fixed or moveable part of a fixture, the purpose of which is to prevent workpiece deflection under the action of imposed cutting forces or clamping forces. Some examples of supports are shown in Figure 2.5.

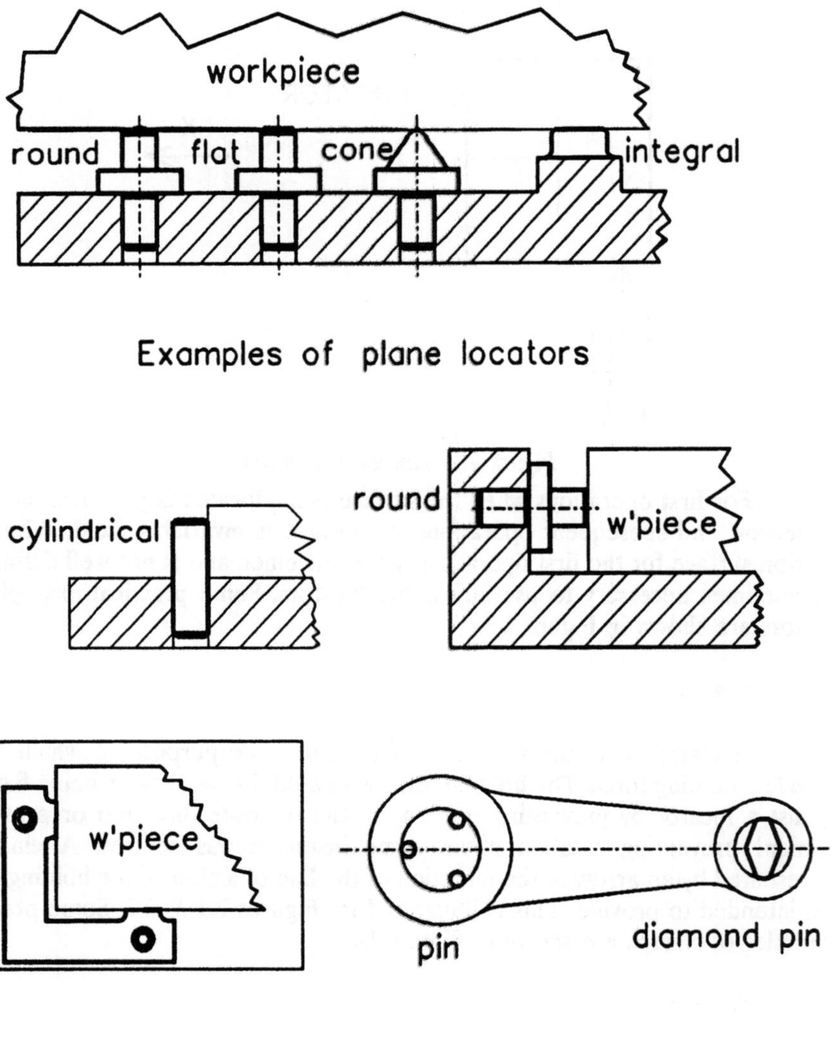

Figure 2.3 Some examples of locators

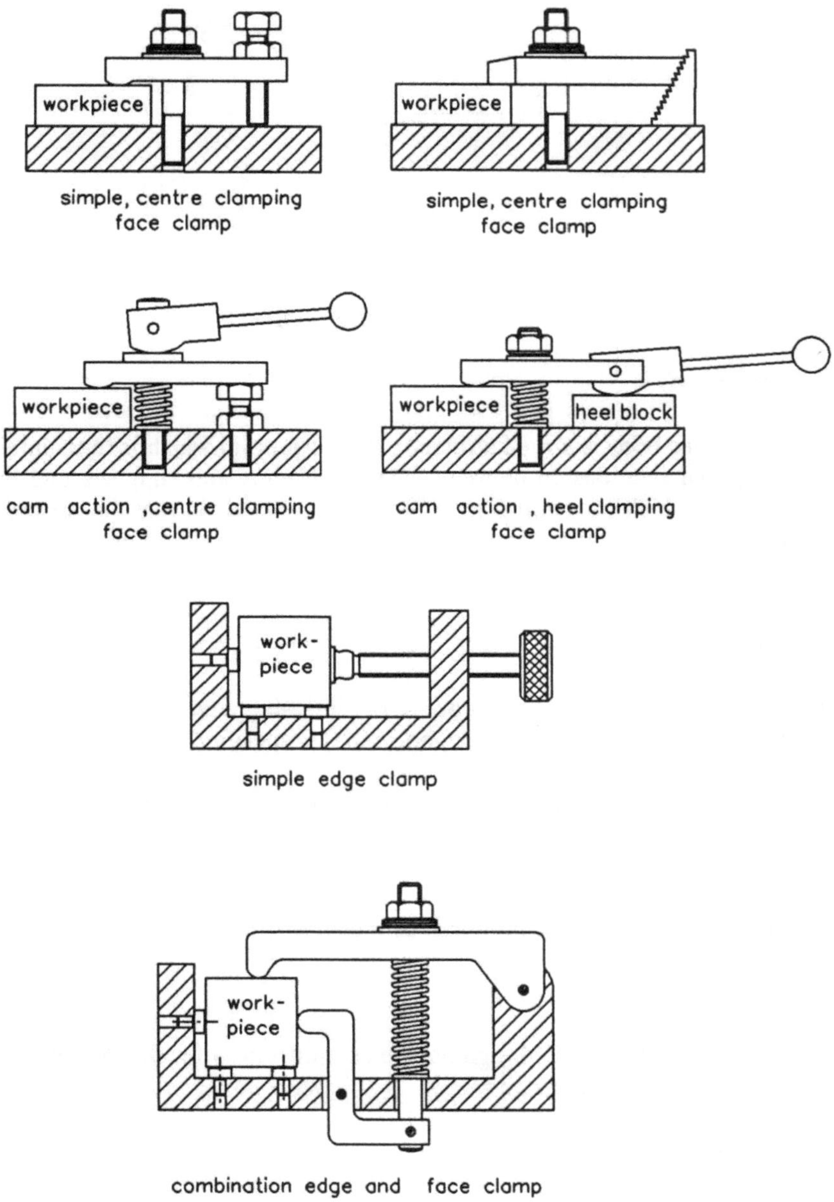

Figure 2.4 Some examples of clamps

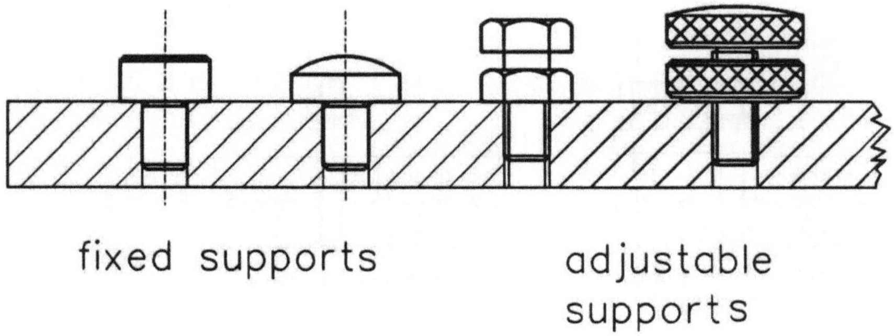

Figure 2.5 Some examples of supports

- *Fixture body*

The fixture body is a rigid structure, the purpose of which is to maintain the correct spatial relationship between locators, clamps and supports and the machine-tool on which the workpiece is to be processed. It may be specially constructed or made up from standard parts as described in Chapter 4.

2.3 Fixture Design

Put simply, fixture design is the process of designing and selecting correct combinations of locators, clamps and supports so that specified design criteria are satisfied.

Fixture design will be considered under the following headings:

design outcomes

- what the fixture design produces as manufacturing information.

design criteria

- the conditions that the design outcomes must satisfy for the design to be considered "a good design."

design techniques

- the techniques employed when developing the design outcomes to ensure that the design criteria are satisfied.

2.4 Design Outcomes

It has been convenient to rationalise fixture design as a number of distinct activities, and variations of the following are found in most published research:

- fixture planning
- fixture layout
- fixture element design
- fixture body design

Fixture planning deals with overall design concepts; fixture layout produces a spatial layout of the fixture; fixture element design is concerned with the details of locators, clamps, and supports; and the fixture body design combines the fixture elements with a supporting structure.

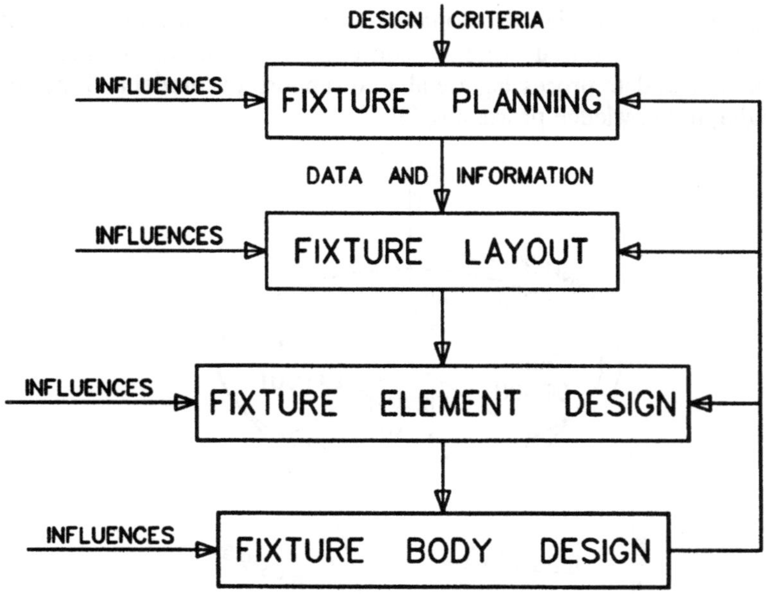

Figure 2.6 Serial model of fixture design

These activities are usually represented as being distinct and occurring serially as shown in Figure 2.6. Because information from later stages will usually have an influence on decisions made in earlier stages, feedback is necessary. The design process becomes a series of nested iterative loops. Such rationalised models of the design process are very convenient and enable a very complex problem to be reduced to a number of bite-sized chunks which can be digested by individual researchers. They also have a structure that is amenable to automation using computers. This however is not how good fixture designers work. The design process is neither serial nor should it be executed as discrete activities. To give a simple illustration: fixture body design is shown as the final event in the design process, however if as a matter of company policy a certain type of modular fixture system is a factory standard, a considerable body of knowledge concerning the fixture body will be known at the start of the design process and this will have significant influence on fixture layout and planning.

The previously defined activities should more correctly be viewed as ***design outcomes.*** What the designer/design process must do is consider the complex field of variables that define the design environment and the relationships between these variables, and assign values and make choices so that the design criteria are satisfied. The outcomes are developed in parallel and not as a series of isolated activities. An alternative model of the design process is suggested in Figure 2.7. This view of the fixture design process is in accordance with modern thinking concerning product design and design of manufacturing processes in general. The prevalent serial/specialist model of the process is being replaced by a parallel model of design development known as concurrent engineering [2]. Doubts have been raised whether conventional CAD techniques based on information and data processing are able to support this model and the emphasis is changing to knowledge processing [3].

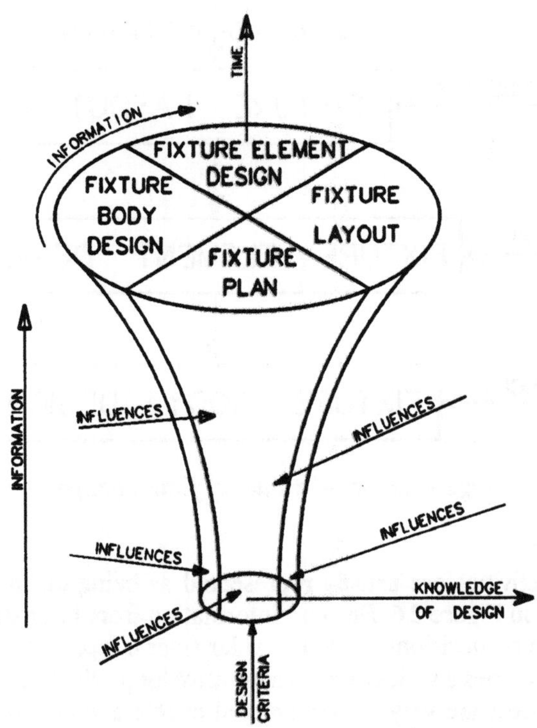

Figure 2.7 Concurrent development of fixture design

Each of the fixture design outcomes will be considered individually together with the factors that influence the design outcomes.

2.4.1 Fixture Plan

The fixture plan establishes the basic fixture concepts. The following outputs are included in the fixture plan:

- **cost analysis**

A cost analysis is required to determine the basic fixture configuration that is economically justifiable.

This is influenced by:

- estimates of manufacturing costs with different levels of automation, labour and machining costs;
- estimates of fixture costs;
- manufacturing methods;
- target batch size and annual production numbers;
- fixture type and complexity;
- number of fixtures.

- **fixture type and complexity**

A decision must be made regarding the type of structure of the fixture and its performance.

This is influenced by:

- factory standards;
- inter- and intra-cell handling in the manufacturing system;
- level of automation;
- technology of the manufacturing process;
- number of operations per fixture;
- cost analysis.

- **number of fixtures (operations)**

The fixture plan must determine whether several operations can be combined in a single fixture or performed in separate fixtures.

This is influenced by:

- component geometry;
- raw material geometry;
- auxiliary processes (hardening, plating, etc);
- orientation of machined surfaces;
- machine tool type and number of controlled axes;
- inter- and intra-machine handling and transport;
- factory standard fixturing requirements.

- **number of fixtures (duplicates)**

It is often necessary to provide duplicate fixtures. This is the case when a manufacturing cell or machine in an FMS has insufficient flexibility to accept different components.

This is influenced by:

- FMS route flexibility;
- FMS mix flexibility;
- FMS volume flexibility.

- **number of workpieces per fixture**

In monolithic FMS with centralised fixturing, it is often necessary to have more than one component mounted on a fixture.

This is influenced by:

- fixture service time;
- inter-cell transport time;
- intra-cell handling;
- machining time per workpiece.

- **orientation of the component with respect to the machine axes**

This is influenced by:

- machine tool configuration (vertical or horizontal spindle);
- cutting technology (end-milling, swarf-cutting, plunge-cutting, etc.);
- number of machine axes;
- position and orientation of location surfaces with respect to the surface being processed;

- weight of the workpiece;
- position of workpiece centre of gravity.

- **location surfaces**

Selection of the surfaces on which the workpiece is to be located is the single most important aspect of fixture design. This decision has a critical effect on the ability of the production process to satisfy the design specifications. Chapter 3 is dedicated to a full discussion of selection of location surfaces.

This is influenced by:

- tolerances on design dimensions;
- form of the raw material;
- sequence of operations;
- orientation of the workpiece with respect to the machine-tool spindle axis;
- spatial relationship of the workpiece, the machine-tool structure and the machining envelope;
- size and shape of available surfaces.

- **clamping surfaces**

Clamping surfaces must be chosen so that all forces imposed during machining can be reacted to the machine-tool bed through the location and support surfaces.

This is influenced by:

- location surfaces;
- support surfaces;
- magnitude and direction of machining forces;
- spatial relationship of the workpiece, the machine-tool structure and the machining envelope;
- surface finishes specified in the design specification;
- workpiece strength and stiffness;
- type and size of clamps.

- **support surfaces**

The need for fixed or adjustable supports must be determined and surfaces for support identified.

This is influenced by:

- location surfaces;
- clamping surfaces;
- tolerances on design dimensions;
- workpiece stiffness;
- magnitude and direction of machining forces;
- spatial relationship of the workpiece, the machine-tool structure and the machining envelope.

- **fixturing sequence**

The above outputs can be combined into a fixturing sequence from which concepts for fixture layouts can be developed.

As can be seen, the fixture plan includes many outputs that are usually considered to be generated by process planning. It is our view that the activities of process planning and fixture design are inseparable, and that practical process plans and fixture designs will only be produced by design systems that combine these activities or develop them concurrently.

2.4.2 Fixture Layout

The fixture layout is an embodiment of the concepts developed in the fixture plan. Included in the fixture layout are:

- **position of locators**

locators must be positioned on the location surfaces so that the workpiece is located in a stable and repeatable manner.

This is influenced by:

- workpiece shape and size;
- surface topology left by previous operations;
- surface deformation experienced by both the workpiece and the locator;
- effect of possible wear of the locators;
- effect of possible build up of dirt and swarf;
- cutting force magnitude and direction;
- shape and size of the locators;
- type of locator.

- **position of clamps**

Clamps must be positioned in such a way that the workpiece is held against all the locators throughout the machining process. Clamps must be outside the machining envelope.

This is influenced by:

- relationship of clamping and location surfaces;
- machining envelope;
- magnitude and direction of machining forces;
- position of locators;
- topography of clamping surface;
- strength and stiffness of workpiece;
- sequence of applying clamps;
- operator convenience and ease of use;
- operator safety;
- swarf clearance;
- cutting fluid application;
- type and size of clamps.

- **position of supports**

Supports must be positioned to minimise deflection and distortion of the workpiece without interfering with accuracy of location.

This is influenced by:

- position of locators;
- strength and stiffness of the workpiece;
- tolerances on design dimensions;
- magnitude and direction of machining forces;
- swarf clearance;
- workpiece topography;
- clamping forces.

- **type of locator**

The decision must be made whether fixed or adjustable locators are used.

This is influenced by:

- accuracy of workpiece blank;
- whether first or subsequent operation;
- use or otherwise of in-process gauging for datum adjustment.

- **type of support**

The decision must be made whether fixed or adjustable supports are used.

This is influenced by:

- tolerance on design dimensions;
- accuracy of workpiece blank;
- strength and stiffness of workpiece;
- automatic or manual fixture loading;
- clamping forces;
- machining forces.

- **type of clamp**

The type of clamp, whether it is to be manually or automatically operated, self-retracting, etc must be specified.

This is influenced by:

- automation techniques to be employed;
- method of loading of fixture;
- size and shape of the workpiece;
- factory standards;
- desired clamping force;
- range of clamp movement;
- machining envelope;
- operator safety;
- ease of use;
- cost analysis;
- swarf clearance;
- clamping sequence.

- **clamping sequence design**

The sequence of applying the clamping force must be determined so that when the clamps are applied, the location surfaces are brought into contact with all of the locators.

This is influenced by:

- position of clamps and locators;
- lines of action of clamping forces;

- relative magnitudes of clamping forces;
- friction at locator/workpiece contact points;
- instantaneous centre of workpiece rotation.

2.4.3 Fixture Element Design

Fixture element designs are detail design drawings of the practical embodiment of the theoretical locators, clamps and supports determined in the fixture plan and fixture layout. Because these elements are well defined and the requirements are common to many different fixtures, it is usually possible to use standard designs or proprietary bought-in components, in which case all that is necessary is to specify components that satisfy the requirements of the fixture layout and plan.

The detail design of all the components of a fixture is influenced by the following factors:

- availability of suitable standard designs;
- number required;
- available manufacturing techniques;
- limitations of manufacturing techniques;
- material properties;
- factory standards.

- **detail design of locators**

  This is influenced by:

  - surface deformation of locator and workpiece;
  - topography of location surface;
  - position and orientation of the workpiece surface;
  - avoidance of wear of the location face;
  - avoidance of build-up of dirt on the location face;
  - provision for replacement of worn parts;
  - adjustment to compensate for wear and manufacturing error;
  - swarf clearance.

- **detail design of clamps**

  This is influenced by:

  - level of automation;
  - method of actuation (eg. mechanical or pneumatic);
  - protection of finished workpiece surfaces;
  - calculated clamping forces;

- required clamp movement;
- operator safety;
- ease of use.

- **detail design of supports**

  This is influenced by:

  - whether fixed or adjustable;
  - method of adjustment;
  - calculated required stiffness;
  - swarf and cutting fluid clearance.

2.4.4 Fixture Body Design

The fixture body provides the structure to combine the fixture elements in the intended spatial relationship with the machine tool bed. The fixture body must satisfy the essentially theoretical requirements specified as outcomes in the fixture layout and fixture plan and in doing so also satisfy an enormous number of practical requirements. The fixture body design will usually be constrained by factory standards which are discussed in the *design criteria*. The following are a selection of some of the factors influencing fixture body design:

- fixture plan;
- fixture layout;
- fixture element design;
- workpiece shape and size;
- machining envelope and cutter access;
- factory standards on construction;
- factory standards on mounting points;
- factory standards on pallet design;
- factory standards on health and safety;
- strength and stiffness of construction;
- machining forces;
- access to workpiece;
- clearance of swarf;
- clearance of cutting fluids;
- application of cutting fluids;
- lifting attachments;
- ease of loading and unloading;
- visibility of locators;
- compatibility with transport and handling equipment.

2.5 Design Criteria

The design outcomes must satisfy several design criteria.

2.5.1 Design Specification

The primary purpose of a fixture is to ensure that workpieces are produced to the design specification. Fixture design can have an effect on dimensional tolerances, geometric tolerances and workpiece surface finish. The ultimate criterion by which a fixture is judged must be its ability to produce workpieces to specification.

2.5.2 Factory Standards

The fixture is part of a much larger production system and its design will invariably be constrained by factory standards which ensure compatibility with the rest of the system.

Typical constraints imposed by factory standards are:

- use of an accepted fixturing system, eg a particular type of modular fixturing system;
- use of a standard baseplate;
- standard tee-bolt spacing;
- standard fixture components;
- level of automation;
- power source;
- control system;
- maximum dimensions to fit available machines;
- maximum weight;
- health and safety standards;
- compatibility and connectivity with handling and transport systems.

2.5.3 Ease of Use

The fixture should be designed to satisfy ergonomic and ease of use criteria. Some of the factors to be considered are:

- health and safety regulations;
- access to clamps;
- minimum use of tools;

- ease of cleaning;
- ease of adjustment and repair;
- size and weight limitations.

2.5.4 Cost

As with any design, an important criterion is that the desired function is achieved at the minimum cost. Cost of fixtures is a major part of the cost of an FMS. Some estimates have placed it as high as one third of the total cost, and any reduction in fixture costs is likely to be significant. Cost of fixtures should not however be considered in isolation from the system as a whole, savings in fixture costs could result in an increase in overall production costs.

2.6 Design Techniques

Various techniques have been employed to guide and assist the designer in generating design outcomes which will satisfy the design criteria, the important techniques are:

- Axiomatic design
- Rule-based design
- Algorithms and analysis tools
- Group technology
- Parametric retrieval
- Design procedures

2.6.1 Axiomatic Design

Axioms are well accepted truths which provide the basis for decision making in the design process. If a design solution is developed applying the appropriate axioms it should follow that the design criteria are satisfied. Eary and Johnson's book *Process Engineering For Manufacturing* [4], introduced the principles of *workpiece control*. These principles form the basis of most fixture design systems and can be treated as axiomatic for good fixture design. Good workpiece control is necessary for the workpiece design specification to be met. These principles are re-introduced here in simplified form and presented in the form of axioms.

Workpiece control

Workpiece control is considered under three headings:

- geometric control
- dimensional control
- mechanical control

• Geometric control

Geometric control is concerned with stability of the workpiece. The position of the workpiece in the fixture is defined by a number of locators. For good geometric control, the workpiece must automatically come into contact with all locators in an exactly repeatable way despite operator skill. Geometric control is based on location theory. Any rigid workpiece has six degrees of freedom and twelve directions of motion. Locators stop movement in one direction only, therefore for complete location exactly six locators are required. Motions in the opposite direction are prevented by holding forces. Consider the cube shown in Figure 2.8, the six degrees of freedom are shown in the isometric view and the six locators necessary to restrain motion in one direction of each degree of freedom are shown in the orthographic views.

▲1 prevents motion in the -z direction

▲2 prevents rotation in the -θy direction

▲3 prevents rotation in the -θx direction

▲4 prevents motion in the -x direction

▲5 prevents rotation in the -θz direction

▲6 prevents motion in the -y direction

Axiom 1 Only six locators are necessary to completely locate any rigid workpiece. Any locators in excess of six are redundant and give rise to uncertainty in location.

Axiom 2 Three locators define a plane.

A corollary of axiom 2 is that it is impossible to simultaneously locate two planes on a workpiece. This is illustrated in Figure 2.9 and correct location is shown in Figure 2.10.

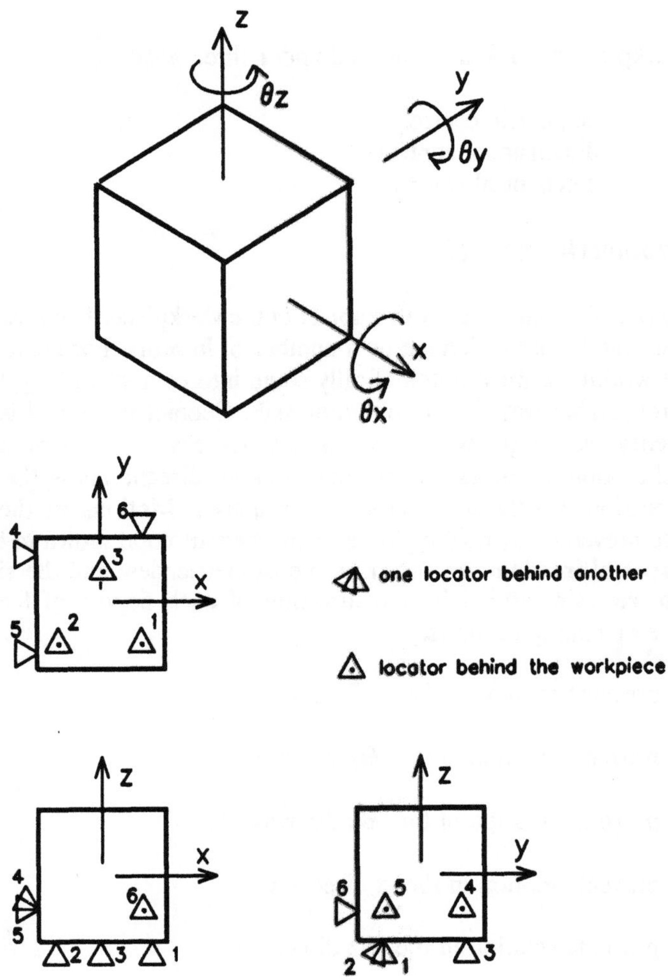

Figure 2.8 The six locators to locate a cube

Axiom 3 Only one direction of each degree of freedom is located.

A corollary of axiom 3 is that locators are not positioned on opposite surfaces.

Axiom 4 Each degree of freedom has only one locator.

Axiom 5 The six locators are positioned as widely spaced as possible to provide maximum workpiece stability and to minimise the effect of wear of locators and workpiece irregularity.

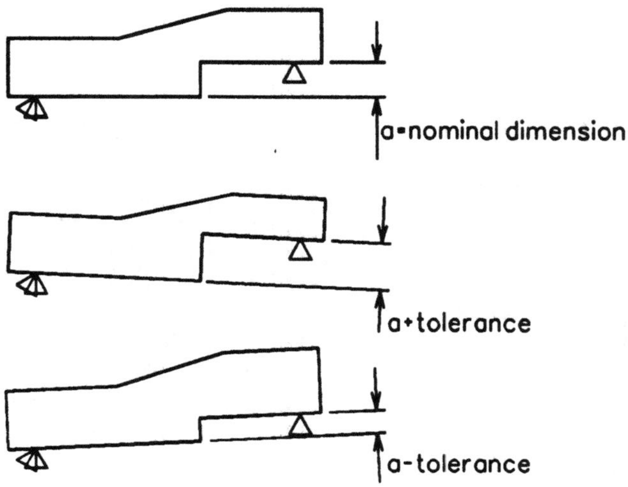

Figure 2.9 *Simultaneous location of two planes showing the effect of variation in dimension a*

These five axioms are usually combined as 'the 3-2-1 location system' which can be used to give correct geometric control for any orthorhombic workpiece. Three locators are placed on the largest planar surface, two locators are placed on the surface perpendicular to the plane of the three locators containing the longest edge, the remaining locator is placed on the mutually orthogonal plane. The 3-2-1 system can also be used to locate the orthogonal datum surfaces of non-prismatic shapes with respect to the principal planes of a machine-tool. An example is shown in Figure 2.11 where the set of two locators is placed on the tangent to two holes.

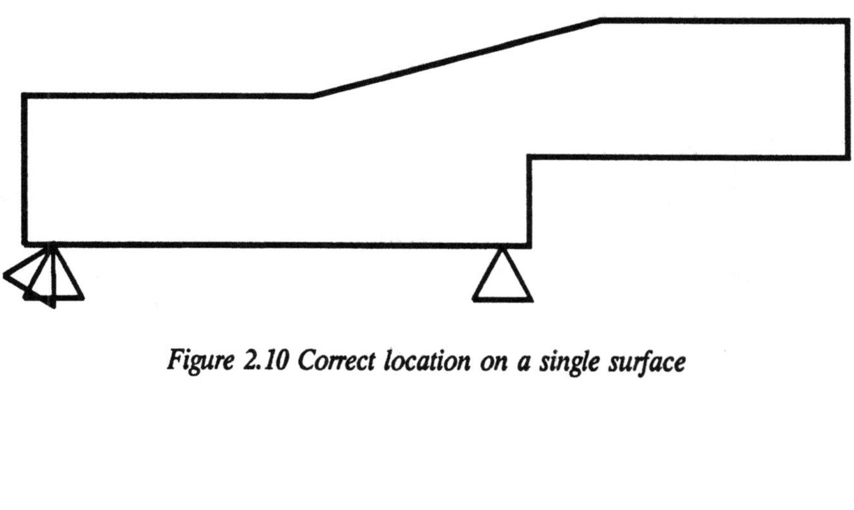

Figure 2.10 Correct location on a single surface

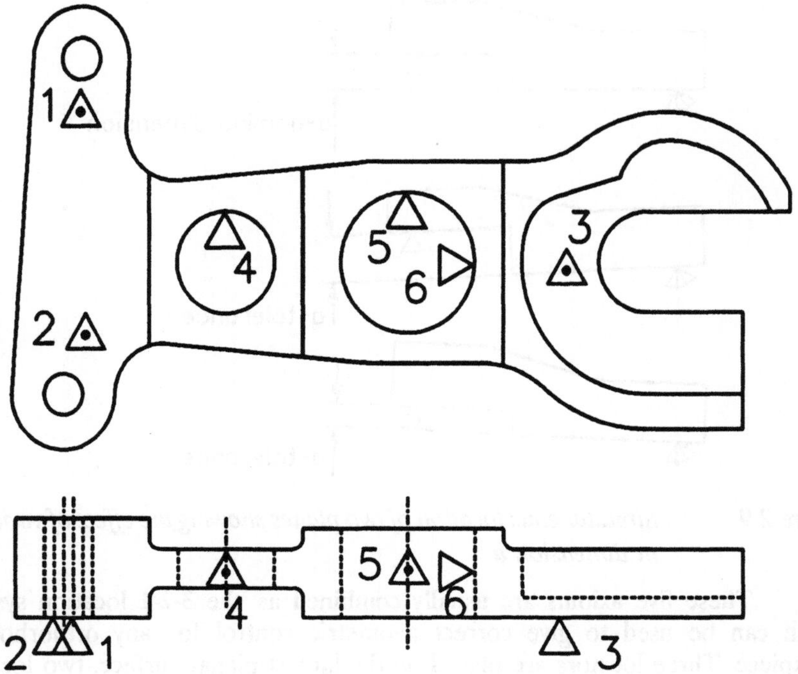

Figure 2.11 Location to two holes

Cylindrical shapes cannot be located using the 3-2-1 system. Figure 2.12 and Figure 2.13 show correct geometric location for long and short cylinders. Considering each degree of freedom in turn will result in the five locators shown in each figure. Because the cylindrical surface is symmetrical about its axis there is no feature of the surface to locate.

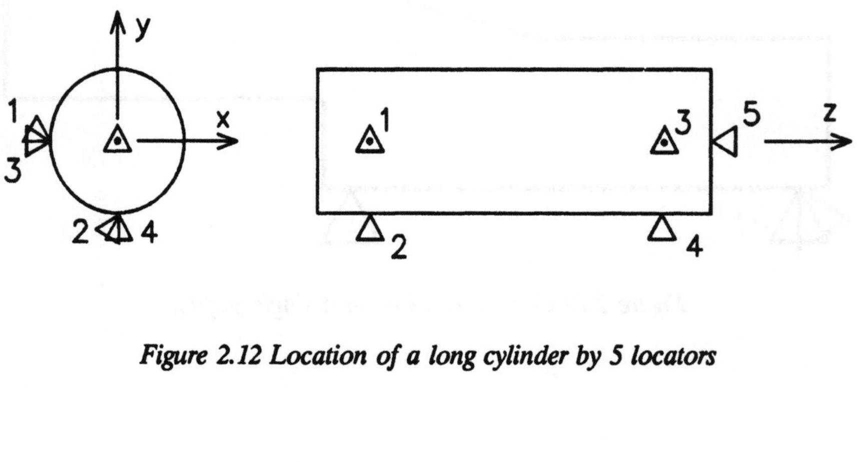

Figure 2.12 Location of a long cylinder by 5 locators

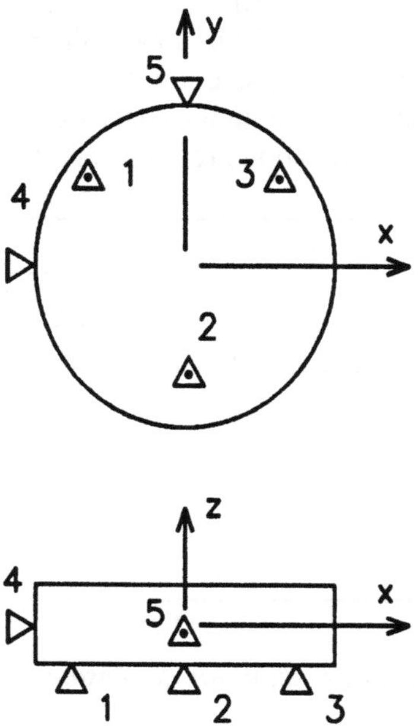

Figure 2.13 Location of a short cylinder by 5 locators

Axiom 6 Only five locators are required to locate a cylinder.

Eary and Johnson have provided many examples of the application of the principles of geometric control to a variety of differently shaped workpieces.The axioms from geometric control are of paramount importance for the fixture plan.

- **Dimensional control**

Dimensional control is concerned with selecting the surfaces for location and position of locators controlling the workpiece position so that the tolerances in the workpiece specification can be achieved and maintained. Good dimensional control exists when there are no tolerance stacks and when workpiece variation and irregularities do not interfere with the correct location of the workpiece. (Tolerance stacks are discussed in Chapter 3.)

Axiom 7 To prevent tolerance stacks locators must be placed on one of the two surfaces which are related by the dimension on the workpiece drawing.

Axiom 8 When two surfaces are related by a geometric tolerance of parallelism or perpendicularity, the reference surface must be located by three locators.

Axiom 8 is illustrated in Figure 2.14. As can be seen, the set of three locators is not on the largest surface. This violates the requirements for geometric control. Operator skill can compensate for poor geometric control but poor dimensional control will always give dimensional variation in the product and in order to satisfy the design specification preference is given to dimensional control.

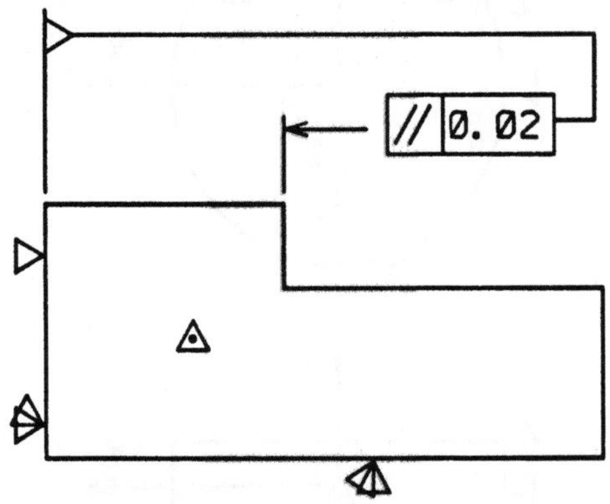

Figure 2.14 Location for dimensional control

Axiom 9 When the requirements of geometric and dimensional control conflict, precedence should be given to dimensional control.

It is common practice to define the position of a surface as a dimension from a centreline of a cylindrical surface, as is shown in Figure 2.15.

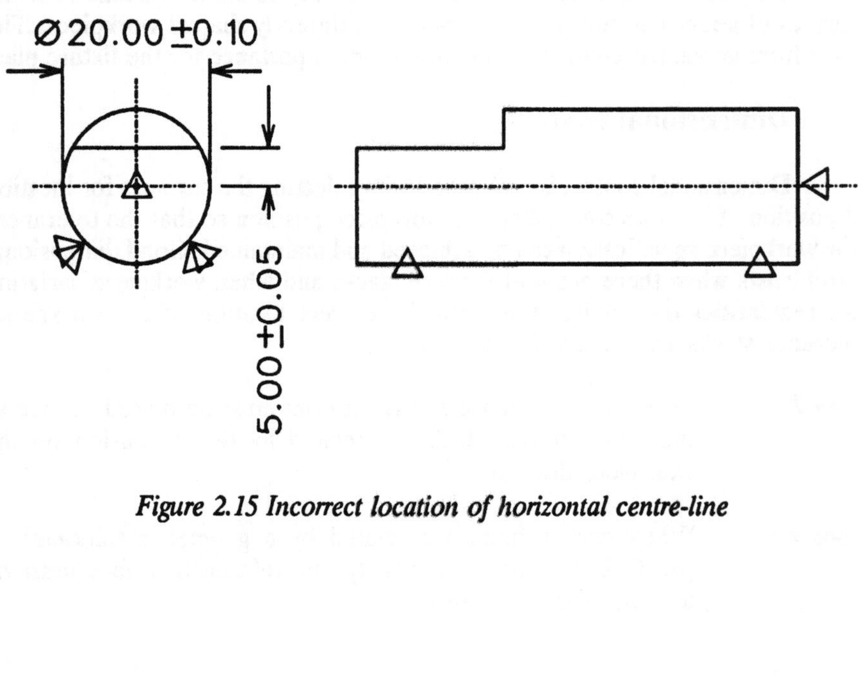

Figure 2.15 Incorrect location of horizontal centre-line

The workpiece is shown located for good geometric control. Located in this way any change in workpiece diameter will cause a change in the position of the horizontal centre-line. The ± 0.1 tolerance on the cylindrical diameter means that the horizontal centreline is only located within ± 0.07mm. Clearly the design specification of ± 0.05 is impossible with this method of location. The locators must straddle the centre-line they locate. An acceptable alternative system of locators is shown in Figure 2.16.

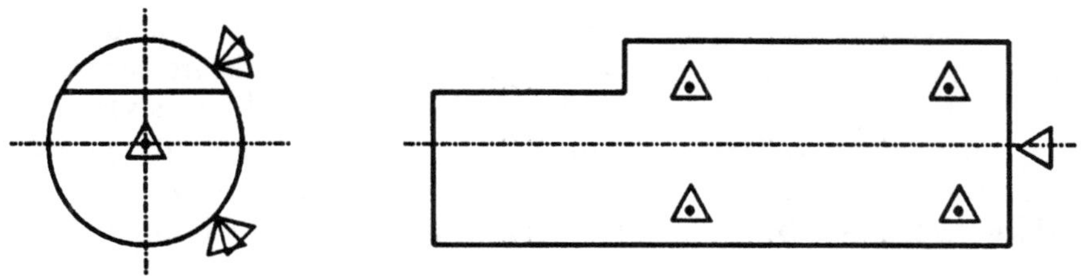

Figure 2.16 Correct location of horizontal centre-line

Axiom 10 To locate a centre-line of a cylindrical surface the locators must straddle the centre-line.

A corollary of this is that it is impossible to simultaneously locate two centre-lines using fixed locators.

The final axiom that can be proposed on the basis of dimensional control is concerned with avoiding the effect of workpiece surface irregularity from unmachined areas such as flash and mould parting-lines.

Axiom 11 Locators should be placed on machined surfaces where there is no possibility of contact with surface irregularities.

The axioms from dimensional control are of importance in producing the fixture plan and the fixture layout.

• Mechanical control

Mechanical control is concerned with control of the effects of cutting forces on the workpiece and the correct placement of holding forces.

Eary and Johnson have identified the following conditions for good mechanical control:

- that the workpiece does not deflect because of the tool-forces;
- that the workpiece does not deflect because of the holding forces;

- that the workpiece does not deflect because of the workpiece's own weight;
- that the workpiece is forced into contact with all of the locators when the holding forces are applied;
- that the workpiece does not shift away from the locators due to the tool forces;
- that the workpiece does not become marred or permanently distorted due to the holding forces.

Good mechanical control is achieved by correct design of the placement of locators, holding forces, (i.e. clamps) and supports in the fixture layout.

All workpieces are elastic and deflect when external forces such as the cutting force, clamping forces, and the workpiece's own weight are applied. In extreme cases the applied force may deflect the workpiece beyond its elastic limit in which case it will be permanently distorted. It is impossible to completely prevent deflection, but the effect on workpiece accuracy can be minimised by minimising the extent of the deflection. This is best achieved by reacting the forces directly to the locators by placing the locators directly opposite the applied forces.

Axiom 12 Place locators directly opposite tool-forces to minimise deflection.

Axiom 13 Place locators directly opposite holding forces to minimise deflection and prevent distortion.

It will usually be a practical impossibility to satisfy both axiom 12 and axiom 13. Axioms 12 and 13 will also frequently conflict with axioms concerning geometric and dimensional control which must be given preference. When axiom 12 and/or axiom 13 cannot be applied, and calculations indicate that deflection will prevent design dimensions being satisfied, then locators must be used.

Axiom 14 If external forces cannot be reacted directly through the locators, then limit deflection and prevent distortion by placing fixed supports opposite the applied force.

Axiom 15 Fixed supports should not contact the workpiece before the load is applied.

Axiom 15 is necessary to ensure that the supports do not prevent the workpiece contacting the locators and in doing so prevent proper location.

When there is uncertainty of the shape of the surface being located or when the design specification requires high accuracy, adjustable supports must be used. Examples of supports are shown in Figure 2.17.

The purpose of holding forces is to prevent the workpiece moving away from the locators. When the holding forces are applied the workpiece should be forced into contact with all locators.

Axiom 16 Holding forces must force the component to contact all locators.

If the strict requirements of workpiece control are followed, six clamping forces are required with a clamping force opposite each locator. Figure 2.18 shows such an arrangement of holding forces.

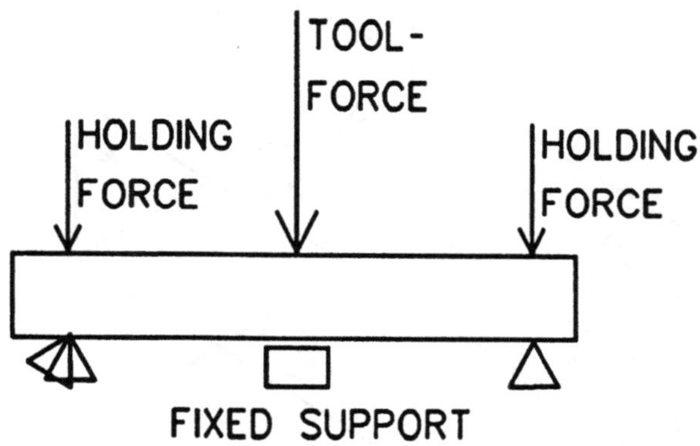

Figure 2.17 Use of a fixed support to resist tool-forces

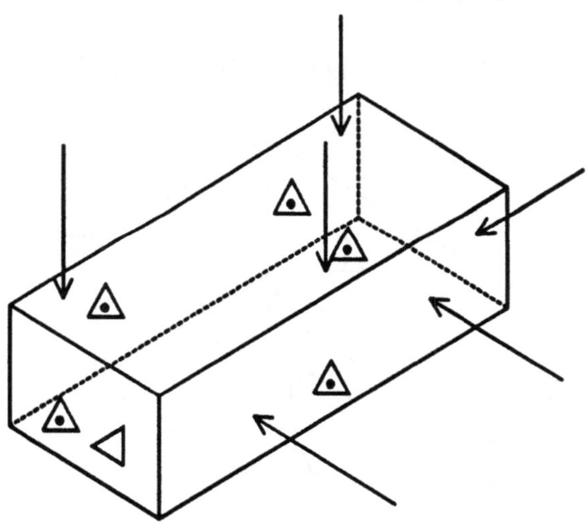

Figure 2.18 Clamping using six holding forces

This is rarely a practical solution, and in any case application of clamping forces would need to be carefully sequenced to satisfy axiom 16. It is usually preferable to reduce the number of holding forces by combining those that are in the same direction. This is illustrated in Figure 2.19. It is possible to combine all three forces into a single force. To satisfy axiom 16 the line of action of the single force must not produce moments about any possible centre of rotation that tend to move the workpiece away from locators. The line of action of the force must therefore pass through the envelope defined by lines joining the three location points locating the base-plane of the workpiece, this is illustrated in Figure 2.20. The component of the line of action of this force projected onto the base plane must also pass through the line joining possible centres of rotation. This is illustrated in Figure 2.21. Referring to Figure 2.21, two conditions are possible.

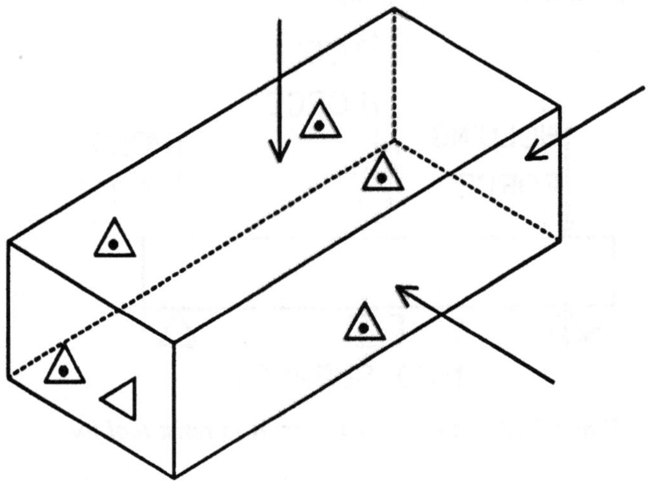

Figure 2.19 Clamping using three holding forces

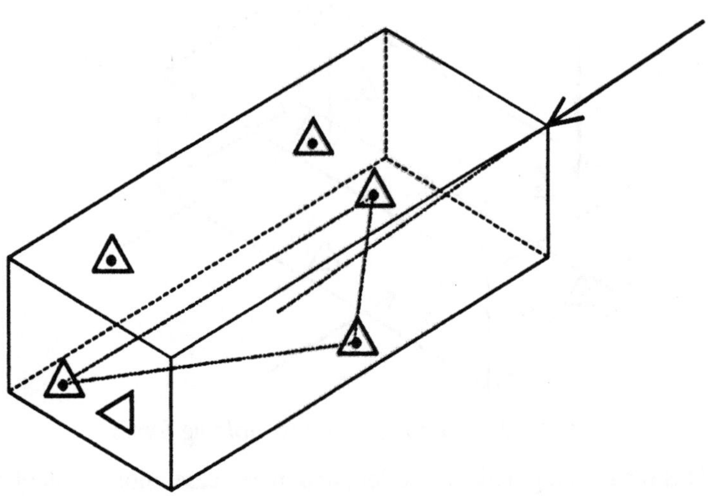

Figure 2.20 Clamping using a single resultant holding force

The first condition is that friction between the locators is sufficient to prevent the workpiece from sliding on the locators. In this case either locator A or C may be a centre of rotation. If C is a centre of rotation the workpiece may, for small movements, rotate about C in a counter-clockwise direction, permitting the workpiece to lose contact at B. The holding force must provide a clockwise moment about C to prevent this. If A is a centre of rotation the workpiece may rotate in a clockwise direction losing contact with B and C. The holding force must provide a counter-clockwise moment about A to prevent this. The workpiece will be prevented from rotating if the holding force passes between locators A and C.

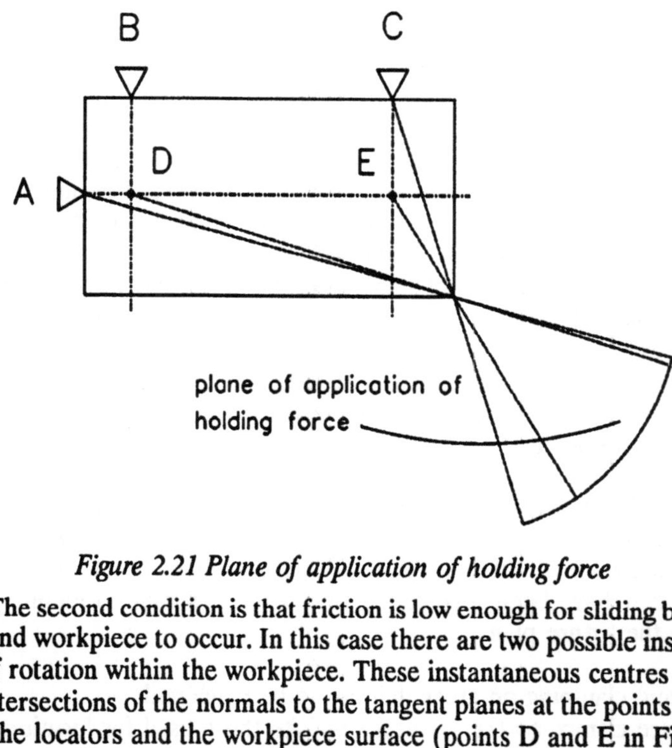

Figure 2.21 Plane of application of holding force

The second condition is that friction is low enough for sliding between the locators and workpiece to occur. In this case there are two possible instantaneous centres of rotation within the workpiece. These instantaneous centres of rotation are the intersections of the normals to the tangent planes at the points of contact between the locators and the workpiece surface (points D and E in Figure 2.21). For small rotations the workpiece may rotate in a clockwise direction about D or in a counter-clockwise direction about E. The moment of the force passing between the centres of rotation will prevent any possible rotation.

When tool-forces are applied, holding forces must overcome any tendency for the tool-forces to move the workpiece away from the locators.

Axiom 17 The moment of the clamping forces about all possible centres of rotation must be sufficient to overcome the effect of tool-forces and prevent the workpiece from moving away from the locators.

If possible, tool-forces should reinforce the clamping forces by forcing the workpiece against the locators.

Axiom 18 Where possible, tool-forces should be such that they force the workpiece into contact with the locators.

A corollary to axioms 16, 17 and 18 is that holding forces should not be in the opposite direction to the tool forces.

These fundamental requirements are axiomatic for good mechanical control.

Other aspects of mechanical control are discussed in Chapter 6.

2.6.2 Rule-Based Design

The 18 axioms are the conditions generally considered necessary for good fixture design. They give little consideration to practical aspects of design and so alone they do not provide sufficient conditions for good fixture design. A convenient way of including practical knowledge in the design process is to use a checklist. Most of the fixture design texts and handbooks listed in the bibliography contain checklists. These checklists are often in the form of questions that the designer should answer when analysing the design at each chronological stage of its development. Checklists of questions may be used in pre-design analysis to ensure that all information governing the design is taken into account. They may be used during the design process to provide critical analysis and after the design has been completed to ensure that nothing has been overlooked.

An alternative form of checklist is a list of guidelines which apply to different design outcomes. Guidelines assist the designer in synthesising new design outcomes. For instance the following guidelines are three of the many guidelines that have been formulated to aid the designer to identify clamping surfaces in the fixture plan.

- avoid clamping on areas that are to be machined
- place clamps on the surfaces opposite those used for location
- avoid clamping on previously machined surfaces

When the design information is in the form of a question it can be inverted to become a guideline. For instance the following questions are from *'A list of questions to be asked before finalisation of the jig design'* in *I.Prod.E Data Memoranda on Jig and Fixture Design* [5], this list has thirty two questions and has relevance to most of the design outcomes in fixture design:

- Are location points clear of flash and burrs?
- Can jig be easily cleared of swarf, particularly the locating surfaces?

The first question can easily be changed into a guideline for fixture layout to become:

- Identify all possible surface irregularities and position locators clear of these areas

The second question is harder to convert into a guideline. To answer this question the designer must know what causes swarf not to be cleared. Consideration of this question gives the following guidelines for fixture body design:

- The locators should be the highest points on any horizontal surface on the fixture.
- The fixture body should be smooth with all screws counter-bored or counter-sunk.

There are dangers in the following guidelines without fully understanding the context for which the guideline was formulated. The most valuable source of design guidelines are books published before robots and automatic loading of fixtures were considered and this should be borne in mind when compiling a list of guidelines. For instance a 4-2-1 location system is often advocated. The reason for this is that if any swarf should adhere to any of the four locators locating the horizontal surface of the workpiece a human operator will be able to detect this by rocking the workpiece on the locators. This departure from good geometric control should not apply to automatic loading of fixtures.

In applying guidelines the designer will frequently find that one guideline will conflict with or contradict another. To resolve such conflicts the design context and the relative importance of every outcome must be understood.

Many of the fixture design systems described in Chapter 5 incorporate the knowledge of these guidelines and dispute resolving procedures as design rules in an expert system to support the design process.

2.6.3 Group Technology

Axiomatic design and guidelines provide a basis for decision making, they cannot themselves synthesise new design solutions. A common approach to design used by many fixture designers is to examine a new fixture problem in terms of the workpiece geometry, the fixture plan, and the technological aspects of the process plan, and by calling on previous experience adapted and combine existing designs to give new design solutions. This approach to design is limited by the knowledge of the individual designer. Group technology provides a structured method of giving a designer access to design knowledge generated previously by other designers. The techniques of group technology have been described in detail in numerous texts [6].

A typical use of group technology is to group workpieces into families on the basis of similarity of shape and production technology. Classification systems are used to describe the features that are used to characterise the workpiece. Classification systems can also be used to retrieve process planning and tooling information for similar parts including information on previously used fixtures. The techniques of group technology have been employed in many computer-aided process planning (CAPP) and computer-aided fixture design systems and have particular use in fixture body design and detail design of the fixture components.

2.6.4 Parametric Retrieval

The techniques of parametric retrieval can be applied to any class of design problem where the item being designed is part of a limited family and where the shape variations within the family can be achieved by simply changing selected dimensions. The relationships between the parameters that define the component shape may be in the form of a simple algorithm. This design technique is particularly powerful when used in conjunction with CAD and many new generation CAD systems, notably ProEngineer and Intergraph's I/EMS II, are based on variational and associative geometry making parametric definition of designs a routine procedure.

Parametric retrieval can be used for detail design of fixture components, locators, clamps and supports and in some circumstances the fixture body. A typical application of parametric retrieval is design of simple components such as clamps. A parametrically defined clamp is retrieved from a library of standard components and modified to suit the particular application.

2.6.5 Algorithms and Analysis Tools

Some parts of the design procedure can be formulated as algorithms and simple equations that can be used to analyse the design outcomes. A small selection of these tools are included here to indicate the range of calculation that is necessary in a complete design procedure.

- **Economic analysis**

The parts of the fixture plan that deal with cost analysis and number and type of fixture are amenable to simple analysis. Equations of the following form appear in many text books:

$$B = \frac{E}{C - D}$$

$$F = A(C\text{-}D)\text{-}E$$

where;

A = Total number of workpieces to be manufactured
B = Economic minimum number of workpieces to justify fixture investment
C = Estimated cost of manufacture per workpiece with simple standard fixtures
D = Estimated cost of manufacture per workpiece with alternative special fixture
E = Estimated cost of manufacture for the special fixture
F = Predicted manufacturing cost benefit from special fixture

Such equations are deceptively simple but should be used with caution. The exact form of the equations depends on company accounting practices. Obtaining accurate values for the cost parameters is likely to be a major difficulty.

• Cutting force calculations

Several design outcomes are influenced by the magnitude and direction of the cutting forces acting on the workpiece and the designer or design system must be provided with a means of predicting these forces. It must be borne in mind that for manufacture using machining centres, a single fixture set-up may be used for a number of different cutting operations using different cutting tools. In a single set-up the workpiece may typically be acted on by drilling, tapping, end-milling and face-milling, each of which will have a quite different resultant effect on the workpiece.

Cutting force prediction has been intensively researched since Taylor's work in 1907 [7]. We owe much of our modern understanding of metal cutting to Merchant and Ernst [8]. More recent work by Oxley and his co-workers has provided a theoretical model that can predict cutting forces in single-point orthogonal cutting with reasonable confidence [9]. Although this model has been extended to include the more complicated cutting geometries found in drilling and milling, the extended models are generally too complex for practical routine design. An alternative approach based on empirical equations or machinability data-bases is favoured. Methods of predicting cutting forces are further discussed in Chapter 6

• Stress and force analysis

The workpiece is an elastic body held in equilibrium in the fixture under the action of the machining forces and the clamping forces. All of the conventional analytical and numerical stress analysis tools may be used to aid the designer to predict workpiece stability and deflection. Some adaptations of these standard techniques applied to fixture design are discussed in Chapter 6.

- **Workpiece control**

Workpiece control has been discussed in terms of axioms, most of which are heuristically derived. An alternative deterministic approach to design of location systems has been proposed by Chou [10]. Chou's method is based on a technique for analysis of freedom and constraint of objects in 3D space known as *Screw Theory* [11]. Chou's method provides an algorithm for synthesis of location and clamping layout design.

- **Tolerance analysis**

Techniques and algorithms for tolerance analysis are discussed in Chapter 3.

2.6.6 Design Procedure

None of the techniques discussed so far provide the designer with a procedure to follow when designing a fixture and in fact this crucial aspect of design has received very little attention in published literature. Most fixture design texts and handbooks consider the various design outputs in isolation with the implicit assumption that the designer will structure their activity in chronological order similar to Figure 2.6. The *SME Handbook of Jig and Fixture Design* [12], suggests an approach summarised in Figure 2.22 which has been reproduced from this handbook. In the SME procedure the design process is divided into five phases each of which is completed before commencing subsequent phases. The different phases relate to design inputs in contrast with the design outcomes which are the concern of Figure 2.6.

Phase one is concerned with all aspects of the workpiece specification, i.e. geometry, surface finishes, tolerances, size, material properties, etc

Phase two is an analysis of the manufacturing operations, i.e. machining, heat-treating, plating, assembly, inspection, etc.

Phase three examines details of each machining operation and the manufacturing sequence and evaluates tentative designs produced during phases one and two in relation to the manufacturing processes. Consideration of workpiece stability and action of cutting forces etc. are introduced here.

Phase four is concerned with operator fatigue, safety and the convenience of using the fixture.

Phase five is the final analysis of the tentative designs and selection of the final fixture design on the basis of overall cost, workpiece quality and production quantity.

An extensive collection of checklists questions the designer on the design inputs. These questions are ordered in such a way that a chronological order is

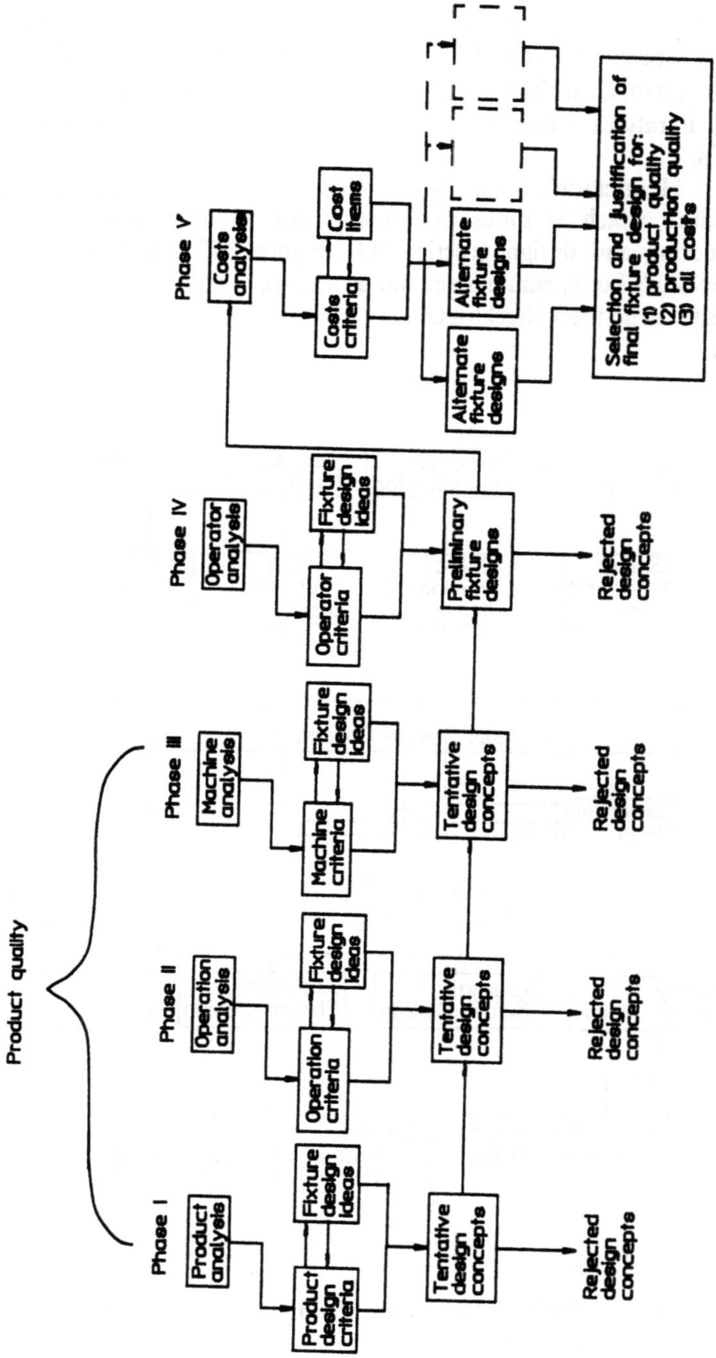

Figure 2.22 *Dependent flow of analytic design criteria, idea and concept variables [12]. (Reprinted with permission of the SME)*

imposed on the designer. During each phase tentative designs are developed which include aspects of all the design outcomes. It should be noted that although the design outcomes are developed concurrently, the design inputs are serial and the design iterations within each phase of the design only consider a limited set of criteria.

Henricksen's book, *Jig and Fixture Design Manual*, offers an alternative procedure which is more in keeping with modern concepts of concurrent engineering and design practice [13]. Figure 2.23 which is reproduced from Henricksen's book, summarises the procedure. The original publication includes extensive structured checklists.

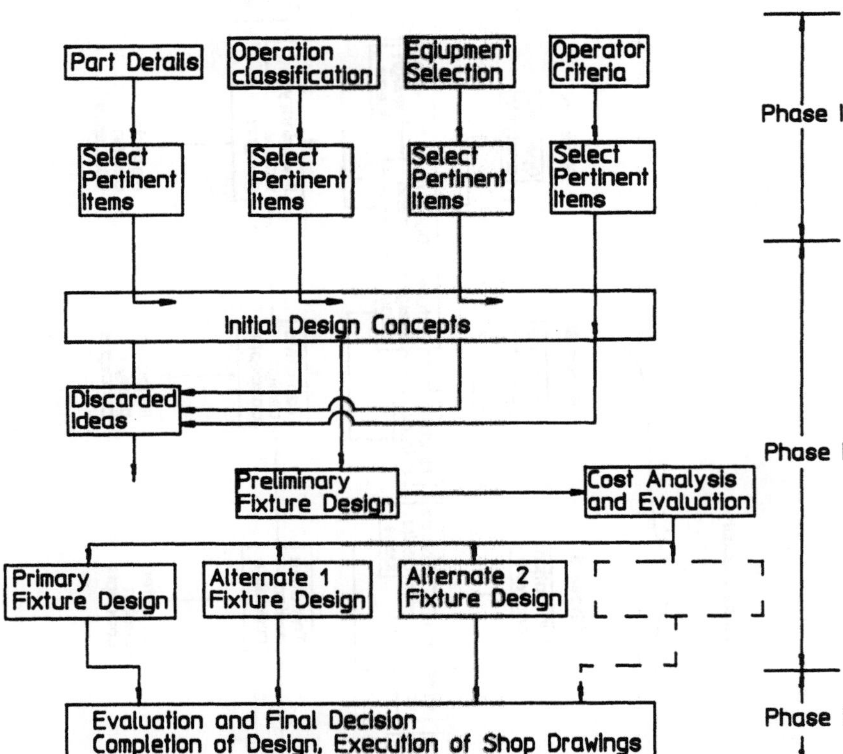

Figure 2.23 *Outline of the fixture planning process [13] (Courtesy of Industrial Press, NY)*

The phases in this diagram are phases of development of design knowledge.

Phase one can be called the information gathering stage. The vertical columns correspond almost exactly with Phases one to four in the SME procedure but here may be executed simultaneously.

Phase two produces a conceptual design based on all aspects of the design specification. A number of embodiment designs are produced and evaluated.

Phase three is the final stage of the design process. The tentative embodiment designs are further evaluated and detail assembly and part drawings produced.

This design procedure has the capacity to progress all design outcomes simultaneously based on the complete design specification.

2.7 Conclusion

A fixture design must satisfy many criteria and is subject to an enormous number of conflicting external and internal influences. None of the above design techniques used in isolation are able to produce practical fixture designs, but when applied in unison they should be capable of forming the basis for a practical computer-aided design method for fixtures. Chapter 5 discusses some of the latest attempts of CAD of fixtures.

2.8 References

1 Coyne RD, Rosenman MA, Radford AD, Balachandran M, Gero JS. Knowledge-based design systems. Addison-Wesley, Reading Mass., 1990

2 Nevins DJ, Whitney DE. Concurrent design of products and processes. McGraw-Hill, New York,1989

3 Lu SCY. Integrated knowledge, geometry, and data processing technologies to support concurrent product development practice. Proc of the Int Conf on Computer Integrated Manufacturing 1991; pp181-184, World Scientific,1991

4 Eary DF, Johnson GE. Process engineering for manufacturing. Prentice-Hall, Englewood Cliffs, NJ, 1962

5 Institution of Production Engineers. Data memoranda on jig and fixture design. The Institution of Production Engineers, London,1973

6 Burbidge JL. The introduction of group technology. Heinemann, London, 1975

7 Taylor FW. On the art of cutting metal. Trans ASME 1907; 28:31

8 Ernst H, Merchant ME. Chip formation, friction and high quality machined surfaces. Surface Treatment of Metals, American Society of Metals 1941; 29: p299

9 Hastings WF, Mathew P, Oxley PLB. A machining theory for predicting chip geometry,cutting forces etc from work material properties and cutting conditions. Proc R Soc London 1980; A371:569-587

10 Chou Y-C. Automatic design of fixtures for machining processes. PhD thesis, Purdue University,1988

11 Ohwovoriole MS, Roth B. An extension of screw theory. J of Mechanical Design, Trans of ASME 1981; 103:725-735

12 Boyes WE (ed). Handbook of jig and fixture design, 2nd edn. Society of Manufacturing Engineers, Dearborn, Mich, 1989

13 Henriksen EK. Jig and fixture design manual. Industrial Press New York 1973

3 Tolerance Control and Location Surfaces

3.1 Manufacturing Tolerance Stacks

A sequence of machining operations involves a series of set-ups for holding and locating the workpiece while it is being machined so that the required design specifications can be economically achieved. To this end the sequence of machining operations and the selection of location surfaces must be based on the practicality of locating and holding the workpiece and on the dimensional relationships between machined features of the part. These are aspects of fixture design previously discussed as workpiece control. Geometric and mechanical control are concerned with practicalities of fixture design and dimensional control deals with the dimensional relationships between machined features. Good dimensional control is guaranteed if the workpiece is located on one of the pair of surface features related by a dimension while the second surface is machined. (This is axiom 7). While this ideal should be strived for, consideration of geometric and mechanical control may make it impossible to achieve. If surfaces not referred to by the dimension are used for location then *manufacturing tolerance stacks* are inevitable.

Figure 3.1 illustrates the simplest example of the occurrence of manufacturing tolerance stacks. The workpiece is completed in a single fixture locating the surface labelled 'A'. The machine tool is assumed to be a CNC vertical spindle mill capable of achieving dimensional accuracy of ±0.05mm. In operation 1 surface C is machined with a working dimension of 25.00mm from location surface A giving a balance dimension AC 25.00±0.05mm. Operation 2 machines surface B with a working dimension 15.00mm from the location surface A giving a balance dimension AB 15.00±0.05mm. The final dimension for AB is the balance dimension AB and satisfies the design specification. The final dimension BC is obtained by subtracting the working dimension AB from AC giving 10.00±0.10 which does not satisfy the design specification. The tolerance on BC is a stack of the tolerances of two other dimensions. Clearly it is not possible to manufacture the workpiece in this way.

In the previous paragraph care was taken to distinguish between different types of dimension relating to the manufacturing process. This distinction is important in tolerance analysis and the different dimensions are defined below.

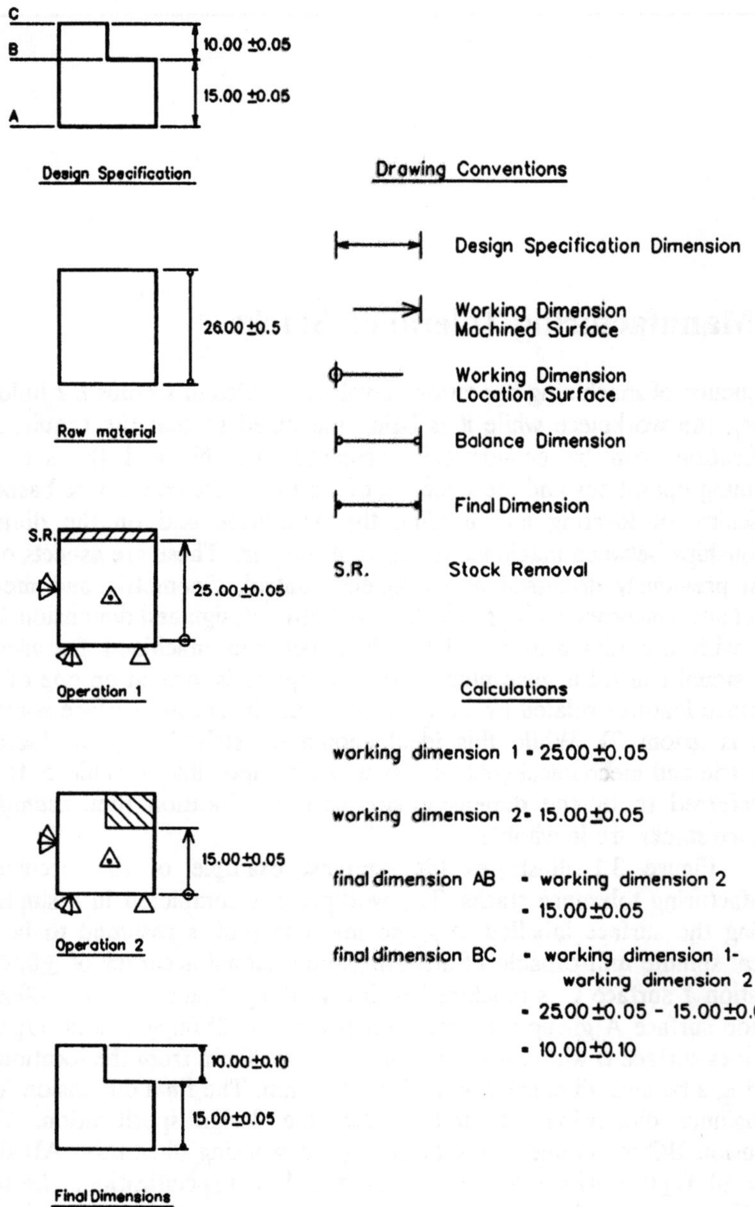

Figure 3.1 Simple occurrence of a tolerance stack

• Design specification dimension

Design specification dimensions are the workpiece dimensions specified in the design drawing. They are sometimes called *blueprint dimensions*. They have two components, a nominal dimension and a tolerance which is the permissible deviation from the nominal dimension. To facilitate calculation it is convenient to convert the dimension so that the tolerance is in equally disposed bilateral form, e.g. dimensions such as:

$$25.00^{+0}_{-0.10}$$

must be converted to:

$$24.95 \pm 0.05$$

In this form the tolerances are always added when a dimension is added or subtracted with another.

• Working dimension

Working dimensions also consist of a nominal dimension and a tolerance on the nominal dimension.Working dimensions have a different meaning in automatic and manual machining operations.

In automatic machining, such as CNC machining, the nominal component of the working dimension is equivalent to the programmed dimension using absolute measurement coordinates. It is the dimension from some specified machine zero to the surface being processed. If the workpiece is accurately located in a fixture, the origin for the dimension can be considered to be the location surface. In automatic machining processes, the tolerance is the accuracy that is achievable by the machine for that particular operation. It is also called *the process capability*. Process capability is usually defined as plus or minus three standard deviations from the nominal dimension.

In manually controlled machining the working dimension has a different meaning. The nominal component is the distance measured from an existing surface to the surface being processed (the existing surface is not necessarily a location surface). The tolerance is an instruction to the machinist of the allowable deviation from the nominal dimension.

• Balance dimension

Balance dimensions are dimensions between workpiece surfaces which can be calculated from working dimensions. If further processing is to be done on either of the two surfaces defining the dimension, the balance dimension is an

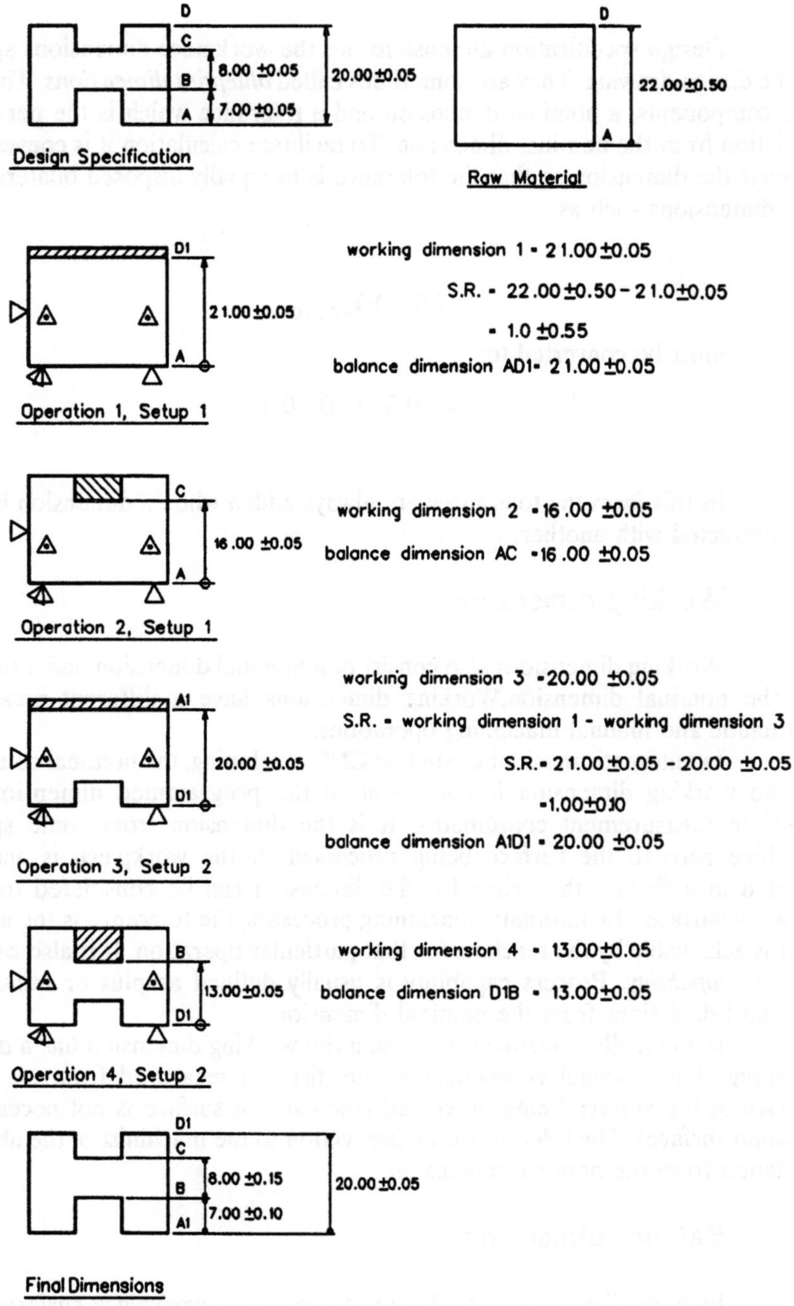

Figure 3.2 Example of two and three high tolerance stacks

intermediate resultant dimension. Because they are calculated from working dimensions, balance dimensions also have nominal and tolerance components. It is usual practice to simply add tolerances when calculating balance dimensions, this gives worst case tolerances and is a very conservative approach. An alternative is to use statistical methods to add the tolerances and maintain the three standard deviation tolerance band. The worst case approach is followed here.

• Final dimension

A final dimension is a special case of a balance dimension when no further processing is done on either surface defining the dimension. For the sequence of operations to be acceptable the nominal component of the final dimension must be equal to the nominal component of the equivalent design specification dimension and the tolerance component must be less than or equal to the tolerance component of the design specification dimension.

Figure 3.1 introduces some drawing conventions which are useful for distinguishing between design specification dimensions, working dimensions, balance dimensions, and final dimensions.

It is very unusual that a workpiece can be completed in one set-up in a single fixture. A workpiece will usually require machining from several different directions requiring different set-ups and fixtures. This will invariably result in higher order manufacturing tolerance stacks. Figure 3.2 shows how tolerance stacks can happen as a result of a sequence of set-ups. The work piece is machined from oversize stock and is machined on all surfaces. As in the previous example a vertical spindle milling machine capable of machining accuracies of $\pm$0.05mm is to be used.

Another diagram label convention is introduced in this diagram, a surface machined more than once is given a subscript for the second and subsequent occurrences of the surface. The subscripts are omitted in the design specification.

Consideration of Figure 3.2 shows that final dimension A1D1 is obtained directly from working dimension D1A1 and hence has a tolerance of $\pm$0.05. Final dimension A1B is calculated from working dimensions D1A1 and D1B, it has a tolerance stack of two and the final dimension is 7.00$\pm$0.10mm.The balance dimension CD1 can be calculated in a similar way to give 5.00$\pm$0.10.

Dimension BC must be calculated from the chain of working dimensions that contribute to the creation of the two surfaces defined by the dimension. The chain of working dimensions is shown in Figure 3.3.

Evaluating this chain:-

final dimension BC = D1B - (AD1 - AC)
where D1B,AD1 and AC are working dimensions

= 13.00$\pm$0.05 - 21.00$\pm$0.05 + 16.00$\pm$0.05

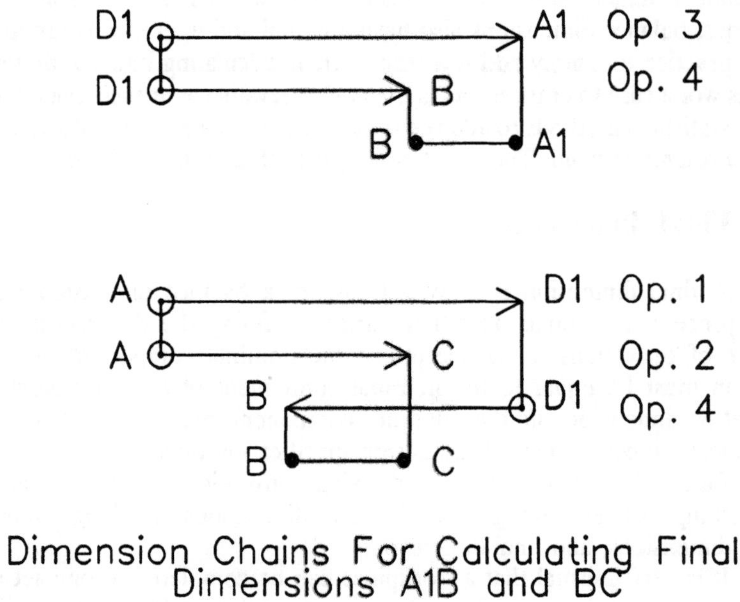

Figure 3.3 Chain for evaluating tolerances

$= 8.00 \pm 0.15$

The dimension BC has a tolerance stack of three and clearly the design specification is not satisfied.

It should be noted that final dimensions must be calculated from working dimensions and not other final and balance dimensions. It might seem that dimension BC could be calculated in the following manner:

final dimension BC = A1D1 - A1B - CD1
where A1D1, A1B and CD1 are final dimensions

$$= 20.00 \pm 0.05 - 7.00 \pm 0.10 - 5.00 \pm 0.10$$

$$= 8.00 \pm 0.25$$

The tolerance calculated in this manner is incorrect as the tolerance on the position of surface A1 is wrongly included and this surface has no effect on dimension BC.

The occurrences of manufacturing tolerance stacks illustrated in Figures 3.1 and 3.2 are as a result of very simple process sequences. In practice the sequence may be much more complex requiring multiple set-ups on several different machines. Higher order tolerance stacks are by no means unusual and they can be quite difficult to detect. Figure 3.4 shows the process pictures for machining a component that requires milling and grinding. It will be shown later that the final dimensions achieved for design dimensions AB and BC are tolerance stacks of four and three other dimensions.

There is a naive though widespread belief that because of the high accuracy of CNC machines, final dimensions with a tolerance similar to the machine-tool accuracy are always guaranteed. Because of the possibility of manufacturing tolerance stacks this is not the case. Process planning systems should minimise the effect of tolerance stacks when planning sequences of operations and selecting location surfaces. Some systematic means of tolerance analysis should be included as an integral part of any practical process planning and fixture design system as a means of checking that the tolerance stacks consequential to the fixturing sequence and location surfaces are such that they do not prevent the design specification being satisfied.

3.2 Tolerance Charts

Tolerance charts have been available for use in process planning in precision manufacturing since the early 1950s [1]. They provide the process planner with a graphical means of displaying all the dimensions and tolerances that contribute to the manufacture of the workpiece at all steps of a processing sequence and a systematic method of tolerance analysis and stock removal verification. They are also an easily understood means for communication between the process planner and product designer. Manufacturing techniques have changed considerably since the 1950s but the importance of tolerance charts has not diminished. Oliver Wade, (a long time advocate of tolerance charting) writing in the fourth edition of *The Tool And Manufacturing Engineers Handbook* [2, p2.1], emphasises the "self evident importance of doing it right first time" and presents tolerance charts as the only systematic way of ensuring tolerance control in manufacture. Unfortunately despite the seemingly obvious enormous benefits from disclosure of tolerance problems before fixtures are built and machining occurs, tolerance charts are not widely used on a routine basis. The reason for this is probably the difficulty and time consuming nature of the 'traditional' methods of chart construction. The majority of computer-aided process planning systems and computer-aided fixture systems are similarly negligent in not giving adequate attention to dimensional tolerance control. A new algorithm for tolerance analysis based on a graph-theoretic approach developed at the University of Canterbury, [3], and [4], is described here. This new method for tolerance charting has benefits for manual tolerance charting but has also been incorporated into a fully automatic tolerance analysis system.

The 'traditional' methods for tolerance chart construction are adequately treated by many other authors, [5], [6] and [7]. The older method will not be described in detail here but is briefly explained with reference to tolerance charts for the sequence of processes in Figures 3.1 and 3.2. The symbols in the tolerance charts are those defined in Figure 3.1.

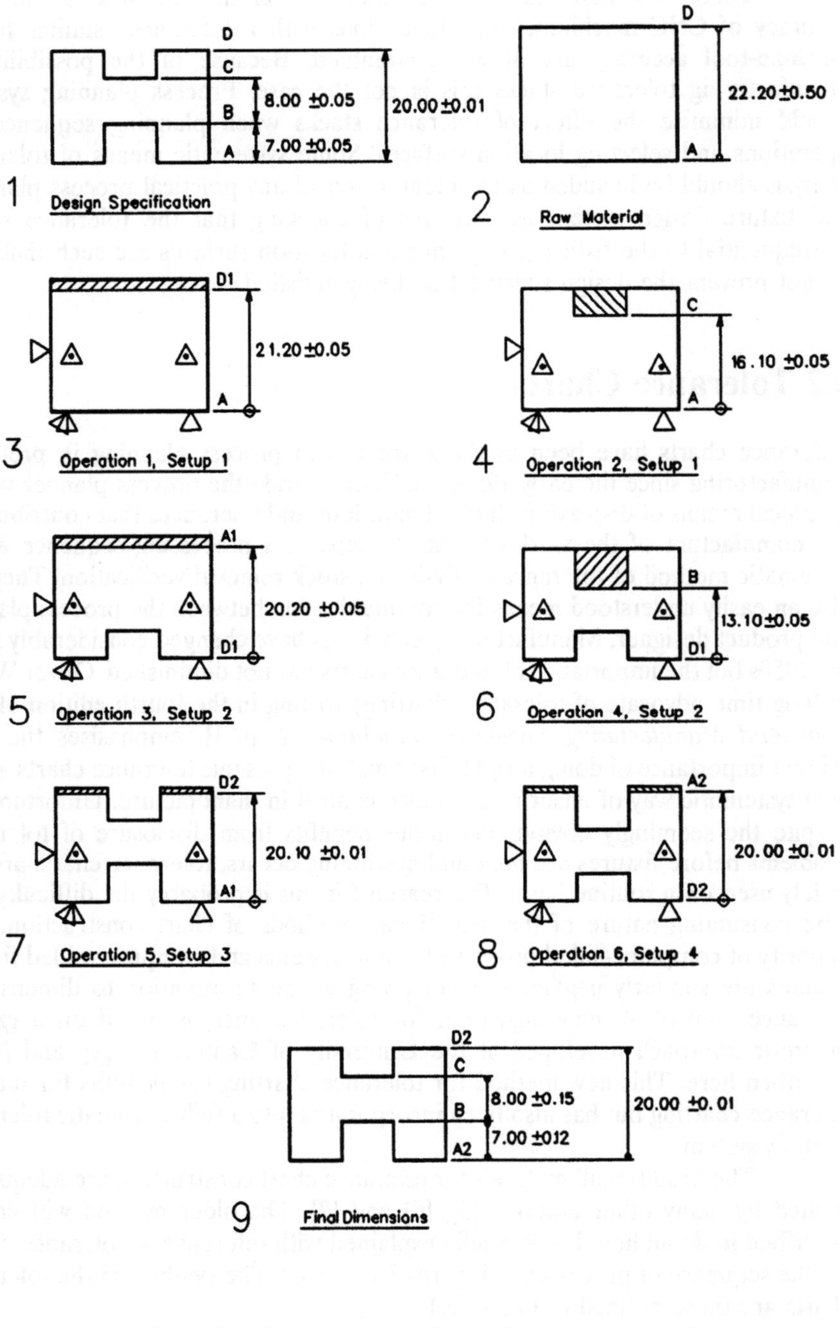

Figure 3.4 Examples of higher order tolerance stacks

1	2	3	4	5	6	7	8	9	10
Line No.	Op. No.	Face	M/c	Working Dim. Nominal	Working Dim. ± Tol.	Balance Dim. Nominal	Balance Dim. ± Tol.	Stock Removal Nominal	Stock Removal ± Tol.
1		A	Reference Surface						
2	Raw	C	saw	26.00	0.50				
3	Mat'l					26.00	0.50		
4	1	C1	mill	25.00	0.05			1.00	0.55
5						25.00	0.05		
6	2	B	mill	15.00	0.05			solid	
7						15.00	0.05		
				Design Spec. Dimension		Final Dimension			
				10.00	0.05	10.00	0.10		
				15.00	0.05	15.00	0.05		

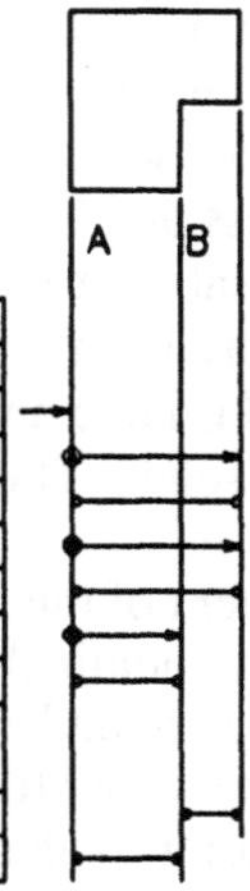

Figure 3.5 Tolerance chart for Figure 3.1

The basic tolerance chart layout is shown in Figure 3.5, this is the chart for the process in Figure 3.1. Leader lines are drawn from a sketch of the surfaces that define the component being machined. The surfaces must be parallel. Separate charts must be constructed for each set of parallel surfaces with different orientations. For instance an orthogonal prismatic component requires three tolerance charts corresponding to the three principle machining directions. Column 1 of the chart is a line number used only for reference purposes. Column 2 is the operation number. Column 3 is the surface being machined, labelled in the manner defined in figure 3.2. Column 4 is the machine being used. Columns 5&6 are the nominal working dimension and process capability. Columns 7&8 are the balance dimensions and the tolerance on the balance dimensions. Column 9 is the stock removal and column 10 is the tolerance on stock removal. The design specification dimensions are entered at the bottom of columns 5&6, and the final dimensions that are calculated when completing the chart are entered at the bottom of columns 7&8.

The starting point for constructing the tolerance chart is a process plan with a sequence of operations and process pictures showing location surfaces, machined surfaces, and stock removal for each operation. At this stage the processes and process limitations and capabilities are known but not the nominal working dimensions and balance dimensions. The process capabilities are the only known tolerances.

The chart is developed by filling in the known information. The chart symbols for working dimensions for each operation and the symbols for the resulting balance dimensions are drawn first. In line 1 surface A is arbitrarily chosen as a reference surface. In line 2 the raw material size is defined as a working dimension, the nominal dimension is as yet unknown but the tolerance is the process capability of the process producing it (in this case a saw). Column 3 can be completed to identify the machined surfaces for each operation. The process capabilities for each operation are entered in column 6 and the nominal stock removals in column 9. The stock removal nominal dimension is a technological property of the process and is usually a minimum depth of cut for the cutting process to perform in a satisfactory way. When the operation modifies an existing surface this value is entered as the nominal stock removal dimension. In the case where the stock removal is large and produces an entirely new surface this is indicated with the word 'solid'. This completes the entry of known information.

The next stage in development of the tolerance chart is calculation of the nominal balance dimensions and the nominal working dimensions. This is done by starting at the bottom of the chart and working back to the raw material dimensions adding on or in some cases subtracting the stock removals for each operation. The balance dimension prior to an operation is the working dimension plus (or minus) the stock removal for that operation. The balance dimension is used to calculate the working dimension higher up the chart. In following this procedure, for most milling operations the stock removal must be added when working back up the chart, that is balance dimensions and working dimensions are larger earlier in the sequence and become smaller as material is removed. There are some exceptions where working dimensions are smaller earlier in the sequence as for example back-boring in milling and right-hand turning on a lathe in which case the stock-removal must be subtracted. Examples are given in Figure 3.6

Applying this procedure to Figure 3.5, in line 7 the nominal balance dimension is the nominal final dimension AB. The nominal working dimension, line 6, must be the same. The nominal balance dimension line 5 is the sum of nominal final dimensions AB and BC1. The nominal working dimension, line 4, must be the same. The nominal balance dimension, line 3, is obtained by adding the nominal stock removal line 4, to the working dimension line 4.

$$1.00 + 25.00 = 26.00$$

This gives the nominal raw material size. At this stage all of the nominal balance dimensions and the working dimensions are complete. When the working dimensions have been determined the tolerances on balance dimensions and stock-removal can be calculated starting from the top of the chart and using the chaining method described previously. The tolerance on the final dimensions then can be compared with the design specification dimensions and it can instantly be seen if (as is the case with all of these examples) the sequence of operations fails to satisfy the design specification.

The tolerance on the stock-removal can be compared with the nominal stock-removal, the minimum stock removal is the difference of these two values. If this value is too small for the process, the stock removal can be increased and

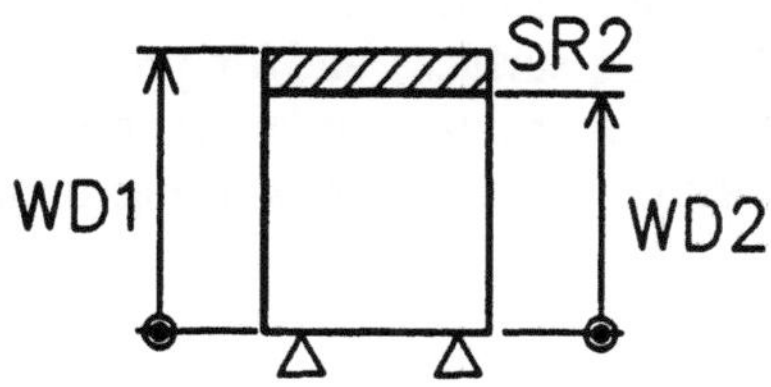

WD1 = WD2+SR

Example 1 Face Milling

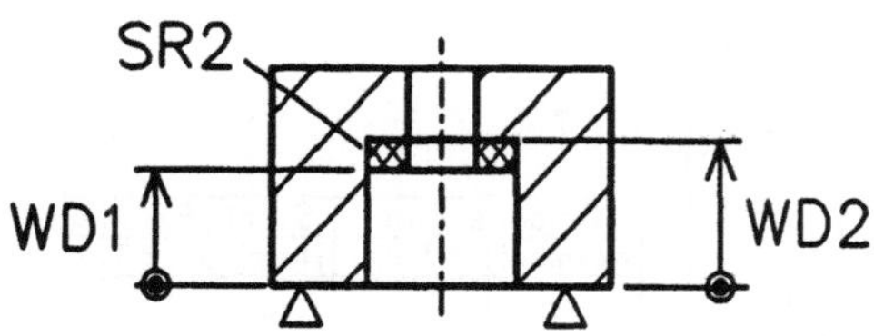

WD1 = WD2-SR2

Example 2 Back Boring

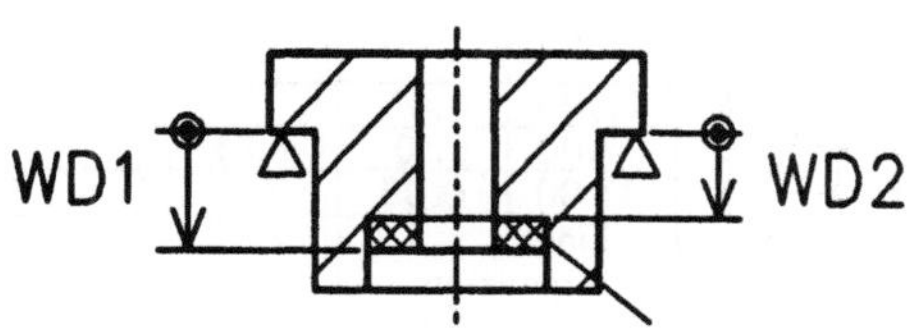

WD1 = WD2+SR

Example 3 Back Boring

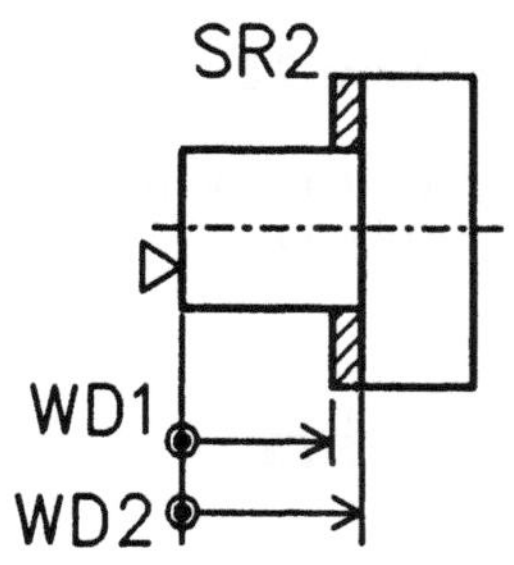

WD1 = WD2-SR2

Example 4 Right-hand Lathe Turning

N.B WD= working dimension
SR= stock removal

Figure 3.6 Some alternative stock removals

the working dimensions re-calculated. If the tolerances on the final dimensions are greater than the tolerances for the design dimensions the process must be re-planned with a different sequence and locating surfaces and/or processes with higher accuracy. If the tolerance on the final dimensions are less than the design dimension tolerances then it may be possible to relax the working dimension tolerances.

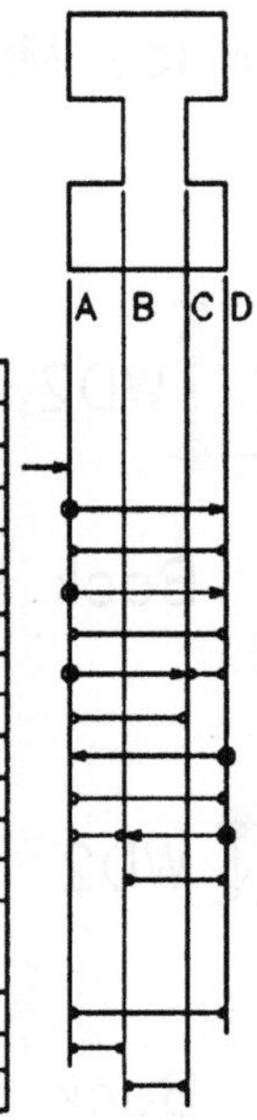

1	2	3	4	5	6	7	8	9	10
Line No.	Op. No.	Face	M/c	Working Dim. Nominal	Working Dim. ± Tol.	Balance Dim. Nominal	Balance Dim. ± Tol.	Stock Removal Nominal	Stock Removal ± Tol.
1		A		Reference Surface					
2	Raw	D		22.00	0.50				
3	Mat'l					22.00	0.50		
4	1	D1		21.00	0.05			1.00	0.55
5						21.00	0.05		
6	2	C		16.00	0.05	5.00	0.10	solid	
7						16.00	0.05	1.00	0.10
8	3	A1		20.00	0.05				
9						20.00	0.05		
10	4	B		13.00	0.05	7.00	0.10	solid	
11						13.00	0.05		

Design Spec. Dimension	
20.00	0.05
7.00	0.05
8.00	0.05

Final Dimension	
20.00	0.05
7.00	0.10
8.00	0.15

Figure 3.7 Tolerance chart for Figure 3.2

Figure 3.7 is the tolerance chart for the process in Figure 3.2. It is a little more complicated but can be constructed in a similar way.

3.3 A New Tolerance Charting Algorithm

It will be realized from following the chart construction for the two simple process sequences considered previously that chart construction for long and complicated process sequences is difficult and time consuming. Moreover the method of construction described in the literature is based on examples and has not been presented as an algorithm or a complete set of construction rules, and as such is not amenable to automation using computers.

Dr.Y.Sermsuti-Anuwat working at the University of Canterbury has developed an automatic fixture sequencing and tolerance analysis program called CAPPFD (Computer-Aided Process Planning and Fixture Design) [8]. This program makes use of a new tolerance charting algorithm based on graph-theoretic techniques. The sequence of operations to produce the component is represented as a ***rooted tree*** which is a special kind of directed graph described by Robinson and Foulds, [9]. A property of the rooted tree directed graph is that the path between each node on the graph is unique. The directed graph can be displayed as a ***tree diagram*** of the kind shown in Figure 3.8 or in the form of a tabulated list. The tabulated list is easily analysed using standard computer techniques but the tree diagram is more suited to manual analysis and it is easier to understand. The algorithm will be explained using the tree diagram representation applied to the sequence of operations shown in Figure 3.4 to develop the chart shown in Figure 3.9.

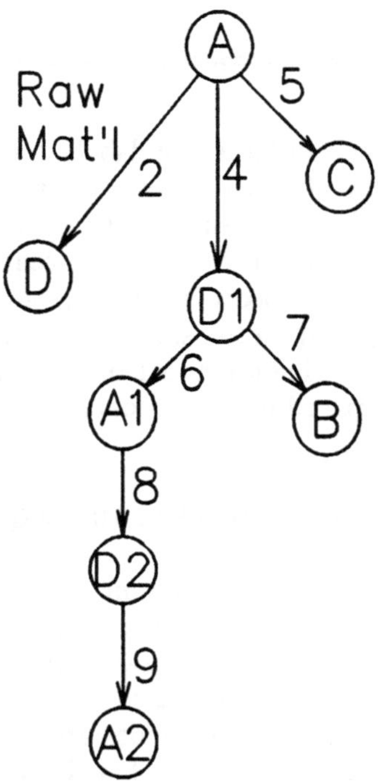

Figure 3.8 Rooted-tree directed graph

• Chart preparation

The initial information is a sequence of operations identifying the locating and machined surfaces. This information is summarised in table 1.

Table 3.1. Machined faces and locating surfaces

Operation	Machined face	Locating face
Raw material	D	A
1	D1	A
2	C	A
3	A1	D1
4	B	D1
5	D2	A1
6	A2	D2

The locating and machined surfaces are labelled in the manner previously described, second instances of a surface are given the subscript #1 and the subscript is incremented for each subsequent instance of the surface. This information can be represented on the tolerance chart in the same way as the traditional method and the process tolerance, stock removal, and design specification columns completed. At this stage all the known information has been entered in the chart. Balance dimensions are omitted as they are not used in the calculations. If they are required for in-process gauging they can be included in the chart.

• Rooted-tree representation of the sequence of operations

The machining sequence can now be represented by the rooted-tree graph shown in Figure 3.8. The surface A on the raw material is chosen as a reference surface and becomes the first location surface. It is the 'root' of the tree and is drawn at the top of the graph. Each node on the graph represents a machined and/or locating surface; most nodes are both machined surfaces and locating surfaces. A link between a pair of nodes on the graph represents a machining operation; the arrow-head points to the surface being machined with the tail end of the link originating at the location surface. The corresponding line number from the tolerance chart is attached to the link. In the rooted-tree graph, the path from one node to another represents the operations that contribute to the distance between the two surfaces represented by the nodes. This path must be used to calculate the dimension between two surfaces. For instance the resultant dimension between A2 and C is the path A2-D2-A1-D1-A-C and the contributing operations are the links on this path, that is operations with line numbers 9,8,6,4 and 5. The tolerance on balance dimension A2C is therefore:

1	2	3	4	5	6	7	8	9	10
Line No.	Op. No.	Face	M/c	Working Dim. Nominal	± Tol.	Balance Dim. Nominal	± Tol.	Stock Removal Nominal	± Tol.
1		A		Reference Surface					
2	Raw	D		22.20	0.50				
3	Mat'l					22.20	0.50		
4	1	D1		21.20	0.05			1.00	0.55
5	2	C		16.10	0.05			solid	
6	3	A1		20.20	0.05			1.00	0.10
7	4	B		13.10	0.05			solid	
8	5	D2		20.10	0.01			0.10	0.06
9	6	A2		20.00	0.01			0.10	0.02
				Design Spec. Dimension		Final Dimension			
				20.00	0.01	20.00	0.01		
				7.00	0.05	7.00	0.12		
				8.00	0.05	8.00	0.15		

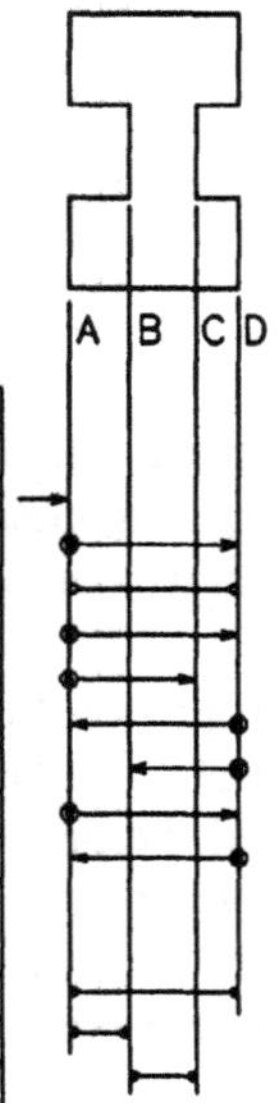

Figure 3.9 Tolerance chart for Figure 3.4

$$0.01+0.01+0.05+0.05+0.05 = 0.17$$

Stock removal is the distance between two surfaces labelled with the same letter but with subscripts differing by one. For example the stock removal for operation 6, (line 9 in the tolerance chart) is the distance A2A1. This is calculated from the operations on the path A2-D2-A1, the contributing operations are line numbers 9 and 8. The tolerance on stock removal is therefore:

$$0.01+0.01 = 0.02$$

• Tolerances on final dimensions

The tolerances on the final dimensions can be calculated by following the above procedure.

The final dimension corresponding to the design dimension AD is A2D2. From the graph, the path is A2-D2 and the only operation is line 9, operation 6. The dimension is given directly by the working dimension therefore the tolerance is 0.01. This value can be entered in the tolerance chart.

The final dimension corresponding to the design dimension AB is A2B. The path is A2-D2-A1-D1-B and the operations are lines 9,8,6 and 7. The tolerance on A2B is the sum of the tolerances for these operations, i.e.:

$$0.01+0.01+0.05+0.05 = 0.12$$

This value can be entered in the tolerance chart. The final dimension corresponding to the design dimension CB is obtained from the path B-D1-A-C and the operations are lines 7,4 and 5. The tolerance on BC is therefore:

$$0.05+0.05+0.05 = 0.15$$

This value can be entered in the tolerance chart. The final dimensions can now be compared with the design dimensions and for this example it can immediately be seen that the design specification cannot be satisfied by this production sequence.

The sequence having been proved unsatisfactory, for practical purposes the chart development would stop at this stage but development of the chart is continued to illustrate the method. If the purpose of the chart is only to check the production sequence no further development is necessary. However the method also provides checking of tolerance on stock removal and working-dimension calculation which are required in a complete process plan.

• Tolerances on stock removal

The method of calculating tolerance on stock removal has been explained. The tolerance on stock removal for operation 6 has already been calculated, the other stock removal tolerances are calculated in a similar way.

The tolerance on the stock removal operation 5 line 8 is obtained from the path D2-A1-D1, the links are the operations 5 and 3, lines 8 and 6. The tolerance on stock removal is therefore:

$$0.01+0.05 = 0.06$$

The tolerance on stock removal operation 3 line 6 is obtained from the path A1-D1-A, the links are operations 3 and 1, lines 6 and 4. The tolerance on stock removal is therefore:

$$0.05+0.05 = 0.10$$

The tolerance on stock removal operation 1, line 4 is obtained from the path D1-A-D, the links are operation 1,line 4 and the raw material dimension, line 2. The tolerance on stock removal is therefore:

$$0.05+0.50 = 0.55$$

Comparing the tolerance on the stock removal with the nominal stock removal, the process planner will see that for the worst case for operation 1 the stock removal is only 0.45mm. From technological considerations such as effect of scale on tool life, this may be deemed insufficient, in which case the process planner may decide to increase the stock removal allowance for this process. This will not effect the sequence of operations or design of the fixtures.

• Unspecified nominal design dimensions

Working dimensions are calculated from nominal design dimensions. Some of the final nominal dimensions are explicitly defined in the design specification but others are implicitly defined and have to be calculated. A matrix method is used for these calculations. The explicit dimensions for this example are shown in bold type in table 3.2

Table 3.2 Matrix for calculating design dimensions

	A	B	C	D
A	0	**7.00**	15.00	**20.00**
B		0	**8.00**	13.00
C			0	5.00
D				0

NB. The suffix notation does not apply in this table

Explicit Design Dimensions

AB = 7.00
AD = 20.00
BC = 8.00

Implicit Design dimensions
(calculated using table 3.2)
AC = AB+BC = 7.00+8.00 = 15.00
BD = AD-AB = 20.00-7.00 = 13.00
CD = AD-AC = 20.00-15.00 = 5.00

• Calculation of working dimensions using the rooted tree graph

The nominal working dimensions can be calculated using the rooted tree graph by working backwards from the final dimensions (the final and design nominal dimensions are the same) and adding or subtracting stock removal with working dimensions.

Rules are required to guide how the calculations are formulated. These rules are based on the topography of the graph e.g. if the path is an alternating set (ie A2-B2-A1-B1) or whether the path follows a fork in the graph. The basic rules were described together with examples in [3] and [4].

The working dimensions are calculated starting at the highest numbered link and applying the following three rules sequentially and repeatedly for each link in the graph.

Rule 1

- The unknown working dimension is equal to the design dimension

This rule applies when no further cuts are made on either of the two surfaces identified by the cut.

In the example being considered this rule applies to A2-D2 and the working dimension for link 9 is the final dimension A2D2

Working dimension, link 9 (WD9) = 20.00

Rule 2

- The unknown working dimension is equal to a known working dimension plus or minus the stock removal, the known dimension and the stock removal are from the next highest link on either of the two surfaces defining the dimension

This rule applies when the surfaces form an alternating sequence or when two cuts are made on a surface using the same locating surface (identified as a fork in the network).

Examples of alternating sequences in Figure 3.8 are A2-D2-A1, D2-A1-D1 and A1-D1-A. Applying rule 2 to these operations:

WD8 = WD9+SR9
= 20.00+0.10 = 20.10

WD6 = WD8+SR8
= 20.10+0.10 = 20.20

WD4 = WD6+SR6
= 20.20+1.00 = 21.20

SR = Stock removal

An example of a fork in Figure 3.8 is D1-A-D. Applying rule 2

$$
\begin{aligned}
WD2 &= WD4+SR4\\
&= 21.20+1.00\\
&= 22.20
\end{aligned}
$$

Rule 3

- The unknown working dimension is calculated from a set of known nominal working dimensions or from a set of known nominal working dimensions and design dimensions.

This rule applies when working dimensions cannot be calculated using rule 1 or rule 2. Calculation of working dimension link 7 (WD7) and working dimension link 5 (WD5) require this rule. From the graph (Figure 3.8), it can be seen that working dimension link 7 is the dimension D1B

$$
\begin{aligned}
D1B &= D2B+D2D1\\
&= D2B+(D1A1\text{-}A1D2)
\end{aligned}
$$

D2B is the implicit nominal design dimension DB from Table 3.2.

$$
\begin{aligned}
WD7 &= DB+(WD6\text{-}WD8)\\
&= 13.00+(20.20\text{-}20.10)\\
&= 13.10
\end{aligned}
$$

Working dimension link 5 is the dimension AC.

$$
\begin{aligned}
AC &= A2C+A2A\\
AC &= A2C+A2A1+A1A\\
&= A2C+(A1D2\text{-}D2A2)+(AD1\text{-}D1A1)
\end{aligned}
$$

A2C is the implicit nominal working dimension AC from Table 3.2.

$$
\begin{aligned}
WD5 &= AC+(WD8\text{-}WD9)+(WD4\text{-}WD6)\\
&= 15.00+(20.10\text{-}20.00)+(21.20\text{-}20.20)\\
&= 16.10
\end{aligned}
$$

In the example being considered the stock removal must be added to a working dimension or final dimension when calculating working dimensions earlier in the sequence. As has been shown earlier this is not always the case and sometimes an earlier working dimension is smaller in which case the stock removal must be subtracted. A simple rule which can be applied if the algorithm is worked manually is that if the dimension corresponding to a working dimension is larger after the cut is made then the stock removal must be subtracted when applying rule 2, if the dimension is smaller then stock removal must be added. This rule is incorporated in CAPPFD together with further rules which determine the effect of the cut on workpiece dimensions.

At this stage all information to complete the tolerance chart has been calculated and the tolerance chart can be completed.

• An alternative method of calculating nominal working dimensions using the tolerance chart

Some exceptions that do not satisfy the three rules for calculating working dimensions from the rooted tree graph have been discovered. These exceptions are not practical sequences and are unlikely to occur in practice. Alternative rules for these exceptions have been formulated, [10], but for complete generality an alternative method based on the tolerance chart and similar to the original manual method is recommended.

The working dimensions are calculated using the tolerance chart by working up the chart and adding or subtracting the stock removals with the design dimensions or the working dimensions. The method is explained with reference to Figure 3.10 which is an enlarged portion of the tolerance chart, Figure 3.9. The explicit and implicit dimensions are required.

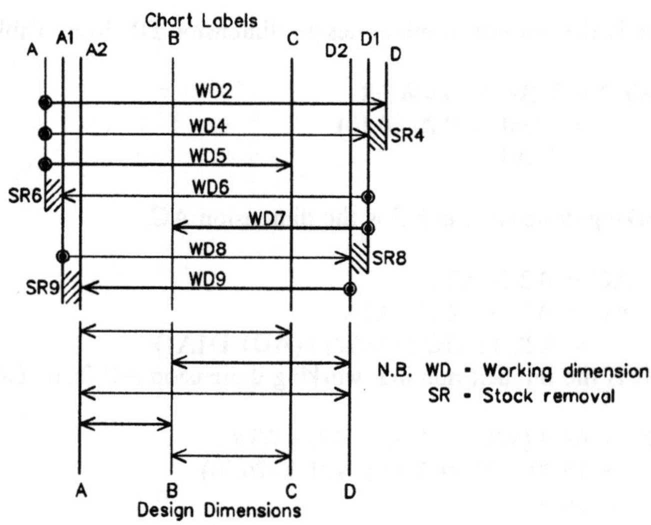

WD9 = A2D2 (or design dimension AD)
WD8 = A2D2+SR9 = WD9+SR9
WD7 = BD2+SR8 (or design dimension BD+SR8)
WD6 = A2D2+SR9+SR8 = WD8+SR8
WD5 = A2C+ SR9+SR6
WD4 = A2D2+SR9+SR8+SR6 = WD6+SR6
WD2 = A2D2+SR9+SR8+SR6+SR4 = WD4+SR4

Figure 3.10 Stock removal calculations from tolerance chart

The method is essentially the same no matter how complex the sequence. Work back from the finished part dimensions which are obtained from the dimension matrix, to the raw material stock dimensions. Start from a final dimension corresponding to the working dimension and add or subtract the stock removal with the working dimension whenever a cut is made on either of the surfaces that define the dimension. If the cut makes the dimension smaller the stock removal is added, if the cut makes the dimension bigger then the stock removal must be subtracted. CAPPFD includes a set of rules to recognise the relationships between surfaces and enables working dimensions to be calculated automatically.

3.4 CAPPFD

CAPPFD is an experimental generative process planning and fixture design program developed to demonstrate the feasibility of automatically generating sequences of machining operations and selecting location surfaces based on the requirements of geometric and dimensional control. The program is fully described in [8]. At present it is capable of sequencing three types of milling operation; plane facing, slotting, and step cutting that can be made on a three axis milling machine. A generic part that can be processed in CAPPFD is shown in Figure 3.11.

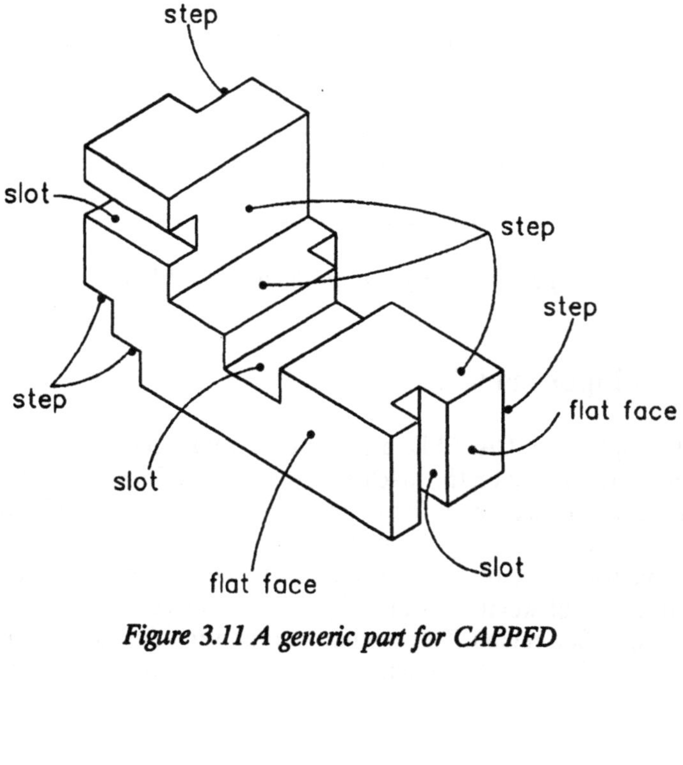

Figure 3.11 A generic part for CAPPFD

The outputs from the system are the sequence of machining operations shown as a set of process drawings similar to those in Figures 3.1, 3.2 and 3.4. These drawings show the condition of the workpiece at the start of the operation and the 3-2-1 location system of location for each operation. A set of three tolerance charts is constructed, one for each principle machining direction.

The CAPPFD system makes use of procedural programming methods and incorporates a decision table rule processor. It is written in C and has been implemented on a 386-PC with 640 kbyte system memory. CAPPFD consists of four main program modules, a support module for producing graphic displays and a utility module to support the calculations. The modules are related as shown in Figure 3.12. The following is an explanation of the program modules.

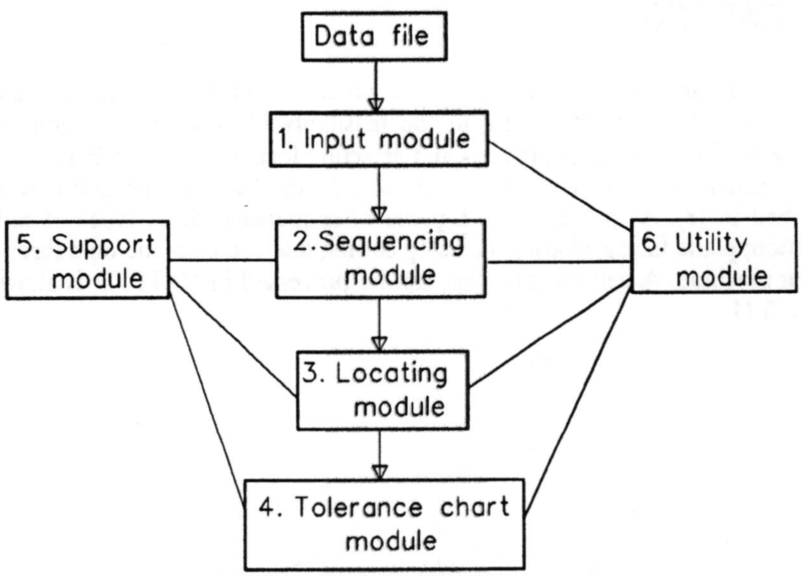

Figure 3.12 Program modules in CAPPFD

• Data input module

The inputs for CAPPFD are a model representing the finished part geometry and information on the machining operations. This information is either retrieved from a prepared data file or through an interactive dialogue with the user. Part geometry is represented as a simple orthorhombic cellular decomposition which enables part geometry to be stored as a series of two-dimensional arrays of numbers which signify the existence or non existence of material in a specified region. This method of representation is described by Ngoi and Whybrew [11]. CAPPFD makes use of this representation because it is computationally efficient and facilitates feature extraction. The other type of data is details of machining operations including number and type of machining operation, the process capabilities, the surfaces constituting the features, and information on pre-formed features such as casting.

• Sequencing module

The sequencing module produces a sequence of operations by taking probabilities of achieving good geometric and dimensional control as the criteria for decisions. The first step is to pre-plan the sequence for different types of operation in the order of the potential of the surface produced to be used for locating the workpiece during other operations. A flat surface has the greatest potential for use as a location surface and hence all flat surfaces are produced first. Steps are the next most useful location surface and are machined next. Slots are of the least use for location and hence are machined last. Having pre-planned the machining order as flat surfaces, steps and then slots the sequence for each feature type is determined based on the degree of dimensional and geometric control that particular feature can offer. This is assessed from the following factors in the following order of importance;

- the number of dimensions on a feature relating to other features,
- the closeness of tolerances on dimensions relating to other features,
- and the degree of geometric control a feature can provide if it is used as a location surface (the larger the area the wider locators can be spaced).

For example, a plane surface to be machined first is the surface that is referenced by the most dimensions; if two surfaces are referenced by the same number of dimensions the priority for machining is based on the closeness of tolerances on the dimensions, the smaller the tolerance the earlier it is to be machined; if the tolerances are equal the surface with the largest area will be machined first. In other words the base lines for dimensions are machined first so that these can then be used for location.

• Locating module

The function of this module is to design the 3-2-1 location system for each operation. The location system is displayed on the computer screen as a process drawing and can be printed out if required. The locating module commences its routines after the machining sequence has been planned by the sequencing module.

CAPPFD uses "backward planning" in designing the location systems, that is it starts by designing the location system for the last operation and then proceeds to the earlier operations. When a location system has been established for machining a particular feature, the information about all locating surfaces on the workpiece is stored and, if necessary, the workpiece geometry modified to be the configuration of the workpiece just before the operation was executed. Therefore the location systems are completely designed when the workpiece geometry becomes the raw material geometry. The process drawings are displayed in the reverse order to the sequence.

When designing a location system CAPPFD first searches for the surfaces on the workpiece that are not on the cutter side. These surfaces qualify as location surfaces because there will not be any interference between location elements of the fixture and the cutting tool. These surfaces are classified as machined and unmachined. The machined surfaces are those having already been machined; unmachined surfaces are either those that do not require machining or are those that require machining but have yet to be machined. CAPPFD stores these two types of surface separately, within each type the surfaces are sorted in a descending order of their areas suitable for locating by three locators. In searching for a set of locating surfaces the machined surfaces are first searched for their suitability for locating by three locators; when found the surface is recorded and the search continued for the surface most suitable for location by two locators; the search is then directed to the surface to be located by one locator. If at any stage the list of machined surfaces is exhausted, the search switches to the list of unmachined surfaces. At each stage the surfaces are considered in order of the suitability of their areas for locating by three locators, that is from large to small.

The criteria for selecting the locating surface(s) for each of the three sets of locators in the 3-2-1 location system are the dimensional relationship and the position of the locating surface(s) relative to the surface being machined. For example, in machining a slot, the surface located by three locators is determined in the following way. If a surface considered is a baseline for a slot dimension, this surface will be selected; if it is not selected, the program will check if the surface is parallel to the bottom face of the slot and if so this surface will be selected; if it is not, the program will continue to check if the surface is parallel to a slot side and if so it will be rejected since it creates a tolerance stack on other dimensions; otherwise it will be selected.

In this way in locating the workpiece, CAPPFD tries to achieve the best geometric control while not jeopardising dimensional control.

• Tolerance chart module

The tolerance chart module in CAPPFD is based on the algorithm previously described. Once the machining sequence and location surfaces have been determined, CAPPFD automatically produces three tolerance charts, one for each of the three principal planes. The operator must inspect the tolerance charts to confirm that the design specification is satisfied. If the design specification is not satisfied the operator must manually edit the process and re-run the tolerance chart module.

3.5 Conclusions

Process planning and fixture design are interdependent and cannot be considered in isolation. Considerations of dimensional control and geometric control are essential if a process plan is to be of practical use. Although these principles are included in manual process planning they are neglected in most computer-aided process planning systems (CAPP) and as such they are unlikely to reliably produce process plans guaranteed to satisfy the design specification. Moreover most of the published descriptions of CAPP do not include any form of tolerance analysis system and are therefore incapable of checking for conformity with the design specification. CAPPFD is an attempt to incorporate process planning and fixture design together and to base the process plan and fixturing sequence on the principles of geometric and dimensional control. It is successful in doing this but CAPPFD is not a complete CAPP or fixture design system; the third component of workpiece control, namely mechanical control is neglected. Chapter 6 discusses some aspects of mechanical control. The criteria used in CAPPFD for determining the sequence of operations and location surfaces are only dimensional and geometric control. Practical considerations of economics that require the number of set-ups and fixtures to be minimised are neglected. CAPPFD makes no attempt at combining operations. These aspects are further discussed in Chapter 5. For these reasons, although the sequences produced by CAPPFD satisfy the design specification, they are usually unacceptable as practical production sequences.

The tolerance charting module has proved to be extremely successful in its own right. Dr. G. A. Britton working at Nanyang Technological University in Singapore, with the collaboration of Sundstand Pacific Pte. Ltd., has developed the tolerance charting algorithm into an independent tolerance analysis and tolerance charting system implemented in Unigraphics UGII. This system called CATCH (an acronym from Computer-Aided Tolerance Charting), provides an interactive user interface and incorporates all the practical requirements for tolerance charting in an industrial environment. Trials at Sundstrand who manufacture precision parts for the aircraft industry have been very successful. Applied to very complex workpieces, some with up to fifty operations, tolerance charts can be produced in under an hour. Using manual techniques it would take a day or more to chart these components.

3.6 References

1 Marks CT. Tolerance charts control production machining. American Machinist March 1953; 97(5):114-116

2 Drozda TJ, Wicks C (ed). Tool and manufacturing engineers handbook, vol 1. Society of Manufacturing Engineers, Dearborn, Mich, 1988.

3 Whybrew K, Britton GA, Robinson DF, Sermsuti-Anuwat Y. A graph-theoretic approach to tolerance charting. International Journal of Advanced Manufacturing Technology May 1990; 5(2):175-183

4 Britton GA, Whybrew K, Sermsuti-Anuwat Y. A manual graph-theoretic method for teaching tolerance charting. International Journal of Mechanical Engineering Education, October 1992; 20(4):273-285.

5 Wade OR. Tolerance control in design and manufacturing. Industrial Press, New York, 1967.

6 Gadzala JL. Dimensional control in precision manufacturing. McGraw-Hill, 1959.

7 Eary DF, Johnson GE. Process engineering for manufacturing. Englewood Cliffs, NJ, Prentice-Hall, 1962.

8 Sermsuti-Anuwat Y. Computer-aided process planning and fixture design (CAPPFD). PhD thesis, University of Canterbury, 1992.

9 Robinson DF, Foulds LR. Digraph: theory and techniques. Gordon and Breach Science Publishers, New York, 1980.

10 Ling KC. Development of computerised tolerance charting system. Project report, School of Mechanical and Production Engineering, Nanyang Technological University, 1992.

11 Ngoi NK, Whybrew K. A fast spatial representation method (applied to fixture design), International Journal of Advanced Manufacturing Technology 1993; 8(2):71-77

4 Flexible Fixture Systems

4.1 Introduction

In Chapter 1 the relationship between fixtures and manufacturing system flexibility was explained. Flexibility in manufacturing systems was discussed in terms of flexibility in scheduling (short-term flexibility), and flexibility in re-configuring the system to accept a not previously manufactured workpiece (reconfigurability or long term flexibility). Manufacturing systems must have both long and short-term flexibility, in order to satisfy both company demands for a high rate of return on investment and market pressures for a short lead-time for production of new products with changed characteristics.

Current FMS demonstrate excellent short-term flexibility but this is achieved by using special purpose dedicated fixtures designed for a single operation on a single workpiece. Such fixtures cannot be used for other workpieces. They are built using mostly non-standard fixture elements and are generally not reusable or reconfigurable. The lead-time to design and manufacture dedicated fixtures is very long. These fixtures increase the cost of production and fixture storage, especially if the parts are in small batches and of large variety. Dedicated fixtures are more appropriate for relatively large production runs requiring high accuracy. This reliance on dedicated fixtures means the FMS have poor reconfigurability.

Fixtures employed in the FMS environment must be readily adaptable. Fixtures must be able to accommodate part families which vary in size and shape and provide quick loading/unloading of the workpiece. Since in FMS the parts are manufactured in batches, the changeover of tools, fixtures and accessories is frequent and the production system must be changed in a very short period of time to keep up with the schedule. To accomplish this, fixtures must be easily modifiable, reconfigurable, disassemblable, attachable and detachable to and from a pallet or machine tool table.

A new class of fixtures called *flexible fixtures* is emerging to satisfy the flexibility requirements of modern manufacturing. The main categories of flexible fixture systems are summarised in Figure 4.6 and will be discussed and evaluated later in this chapter. Previous reviews of flexible fixtures have been published by Thompson [1], and [2].

4.2 Evaluation of Fixture Systems

Fixture systems, like any other component of a manufacturing system, should be evaluated on the basis of *cost, fitness for purpose* and in the present context *long and-short-term flexibility.* Dedicated fixtures will first be assessed on the basis of these criteria and compared with flexible systems in general. In later sections the same evaluation criteria will be applied to specific flexible fixture systems where information is available.

4.2.1 Capital Cost

Capital costs include the following:

- *Design costs*

Design costs for dedicated fixtures are high. The designer has an almost infinite range of possible design choices. Given such freedom a designer can arrive at optimum design solutions but the time involved is long and hence expensive. Flexible fixtures, on the other hand, must follow a prescribed, limited set of design solutions. Satisfactory, though not necessarily optimal solutions can be designed quickly and comparatively cheaply.

- *Fabrication costs*

Fabrication costs include the cost of materials, bought in components, machining, heat treating, finishing, and assembly. All these costs are expensive for dedicated fixtures because they are produced in a toolmaking environment that does not afford the economies of scale that flexible fixtures produced for a larger market can claim. Assembly is particularly expensive for dedicated fixtures because components must be fitted to achieve design tolerances.

- *Commissioning costs*

Commissioning costs are the costs involved in trying out a new fixture, inspecting it to ensure it meets design specifications and making any adjustments found to be necessary. Because of their inherent unique design, these costs are high for dedicated fixtures. Commissioning costs for flexible fixtures are much lower in comparison because they follow a common design philosophy and hence have a more predictable performance.

- *Amortization factor*

Direct comparison of capital costs for dedicated fixtures and flexible fixtures is difficult. The capital cost of dedicated fixtures must be amortized over the production of a single component, whereas the cost of the elements of a flexible fixture system can be recovered from the manufacture of a range of parts over an extended period of time. For a fixture constructed using flexible fixture elements, the amortization factor is the capital cost of the elements used in the fixture divided by the total number of other (different) parts that may use the same fixture elements. The amortization factor should be used when comparing capital costs of various systems.

4.2.2 Recurring Costs

The following recurring costs must be considered when comparing fixture systems:

- *Set-up and stripdown costs*

This is the cost of the time spent in changing from one scheduled workpiece to another. It is directly related to short-term flexibility. With properly designed dedicated fixtures these costs can be very low. The fixtures have a structure independent of the machine tools and pallets and require a minimum of resetting when they are mounted on a pallet. Generally they are provided with features to allow quick mounting and alignment with the machine-tool's principal axes and all that is necessary is to re-establish datum settings for numerical control programs. This is often done automatically with touch sensors and automatic datum correction. Flexible fixture systems, on the other hand, will often have to be reconfigured to accept a change of workpiece.

- *Reconfiguration cost*

With some flexible fixtures, such as modular fixture systems, on the completion of a scheduled batch of components, the decision must be made whether or not to dismantle the fixture or store the fixture complete. The decision is dependent on the frequency with which batches are repeated. If the fixture is stripped, the next time the workpiece is scheduled there will be a reconfiguration cost as a result of the time spent in re-building the fixture. If the fixture is stored intact then the amortization factor will be higher because the fixture components will not be available for use in other fixtures.

- *Maintenance cost*

With use, location surfaces will wear. Dedicated fixtures should be checked periodically and adjustments made to bring them back to specification.

If the fixture is properly designed with adjustable or replaceable location elements, this need not be an expensive exercise. If on the other hand, the fixture is made with solid integral locators, adjustment can be difficult and expensive. Flexible fixtures are not usually used long enough in a single batch to suffer significant wear before being reconfigured for a new part. Adjustments are usually part of the process of reconfiguration. Wear of modular fixture components can be difficult to detect and can make repeatable reconfiguration difficult.

- *Storage costs*

Dedicated fixtures must be stored between batches. With a wide range of products the number of fixtures in storage can be enormous, resulting in high warehousing costs.

4.2.3 Fitness for Purpose

The fitness for purpose of a fixture system when used in the manufacture of a particular workpiece must be judged on the basis of success in satisfying the design criteria discussed in Chapter 2, section 2.5. These criteria are:

- Ability to produce workpieces consistent with the design specification.
- Conformance to factory standards
- Ease of use
- Cost

Special purpose fixtures will always out-perform flexible fixtures on the basis of the first three criteria because they can be specially designed and optimised to satisfy these criteria without compromise to allow reconfigurability. It is in the last criterion, that of cost, where flexible fixtures have the advantage. Nevertheless to be acceptable, flexible fixtures must satisfy minimum standards for the first three criteria even though they may not be an optimal solution. Some general problems and practical considerations when using flexible fixtures will be discussed.

To satisfy the design specification, the axioms and guidelines discussed in Chapter 2 must be followed. Generally the basic fixture functions of location and support must be provided. Flexible fixtures will usually have components which correspond to locators, clamps, and supports and a fixture-body to maintain the correct spatial relationship of these components.

- *location*

Many of the flexible fixtures discussed in section 4.3 provide inadequate or no location, they only provide support. This is not a denial of the basic principles of fixtures because the location function can often be separated from the support function. In the case of phase-change fixtures, for instance, where the workpiece is encapsulated in a matrix of supporting material, location is provided by a separate jig which positions the workpiece while the matrix solidifies. With CNC machines with touch sensing and automatic datum correction, the location requirement can be relaxed. It must be appreciated that with a three axes milling machine, datum correction can only compensate for errors in position, five axes are required to compensate for errors in orientation of the workpiece.

- *supports*

Different workpiece geometries require specialised forms of support. For instance Figure 4.1 shows typical methods of supporting planar and cylindrical surfaces (N.B locators have been omitted in this diagram). Sculptured or freeform surfaces require special support, and considerable ingenuity has resulted in the development of fixture systems such as conformable clamps and phase change fixtures to meet these needs. Many of the flexible systems are limited in the type of surface they can support, and the type of system must be matched with the workpiece family geometry.

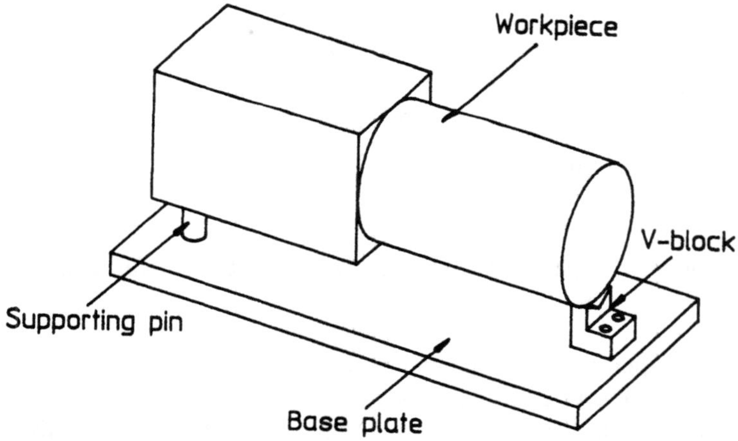

Figure 4.1 Effect of geometry of workpiece on supporting elements

- *clamps*

Clamps play an important part in determining the performance of a fixture. They must satisfy the design axioms relating to mechanical control, that is they must ensure that the workpiece is maintained in contact with locators and not moved away from the locators by tool forces. They must also satisfy practical requirements of ease and speed of use, allow fast loading/unloading of workpieces, and not interfere with the cutting tools' operating envelopes. There are obvious advantages in making clamps completely automatic, especially when used with robot workpiece loading. In general, side clamps provide easier loading/unloading of the workpiece than top clamps as the former are relatively easy to de-activate. Moreover, the de-activated top clamp must be turned away from the workpiece in order to provide a clear space for loading/unloading as shown in Figure 4.2. Top clamps invariably present problems of interference with

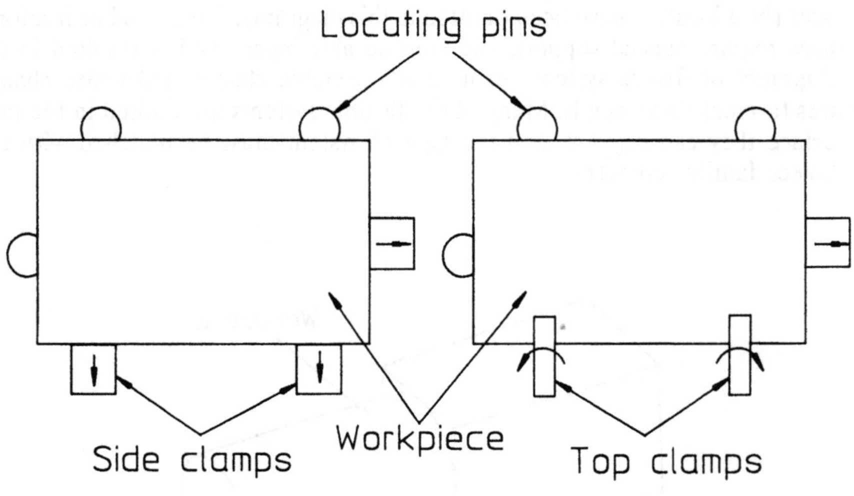

Figure 4.2 Side and top clamps showing their clamping actions

the cutting-tool operating envelope and care must be exercised to avoid expensive collisions.(This is illustrated in Figure 4.3.) Hence, there are clear advantages in the use of fast acting, low profile, side clamps and they are frequently used in dedicated as well as flexible fixtures. Side clamps have an intrinsic problem which is illustrated in Figure 4.4. When the clamping force is applied, the body of the clamp is deformed resulting in a force component away from the base locators in violation of axiom 16. This condition is commonly experienced when workpieces are clamped in machine vices and skilled operators will usually beat the workpiece with a lead mallet to bring it back in contact with the bottom locators (usually the bed of the vice). This is clearly not a desirable procedure for an FMS and several flexible pallet systems which incorporate automatic hydraulic clamps solve this problem in the manner shown in Figure 4.5. By using twin angled clamps the horizontal forces on the assembly are balanced and the clamps provide vertical force components towards the base locators. These are called ***vectored clamps***. Such features are difficult and expensive to incorporate in dedicated fixtures and are more commonly found in commercially available flexible fixture systems [3].

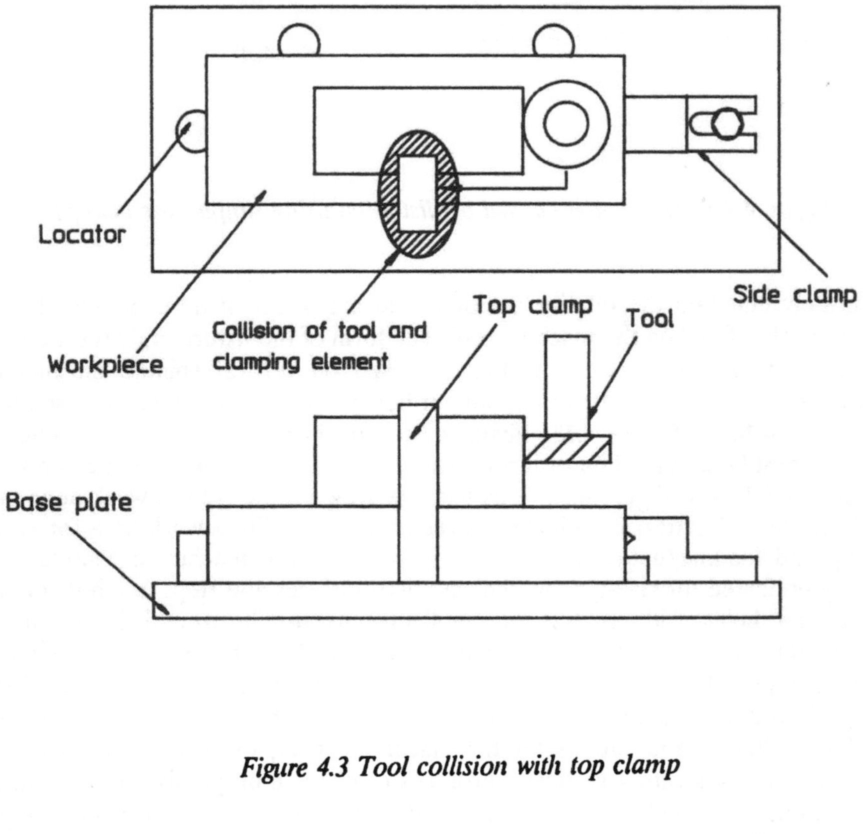

Figure 4.3 Tool collision with top clamp

- *fixture body*

The fixture body design is largely responsible for the short term flexibility of any fixture system. It is the fixture body which is attached to the machine-tool or pallet and for maximum flexibility must be capable of rapid removal replacement and re-alignment with the machine tool axes. The fixture body must

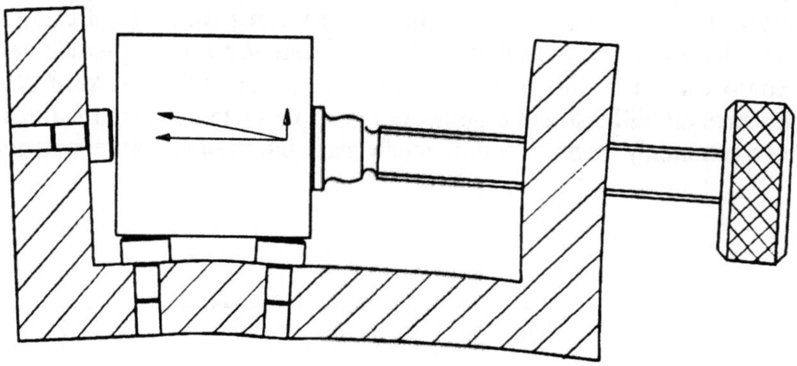

Figure 4.4 Poor location caused by distortion using simple side clamps

accurately maintain the relative positions and alignment of all elements of the fixture both off and on the machine-tool. The form of the fixture body is different for different machine-tool type. Fixtures used on vertical spindle machining centres must incorporate or be capable of being mounted on baseplates, similar to Figure 4.10 or Figure 4.11. Fixtures used on horizontal spindle machining centres must be mounted on or incorporate tooling blocks or angle-plates similar to Figures 4.12 and Figure 4.13. This has a bearing on the type of work done on each machine. Vertical machining centres provide a defined work area for easy set-up and loading/unloading of the workpiece. These machining centres are usually preferred for slab-type workpieces with surfaces and steps in a horizontal plane and holes with centres perpendicular to the horizontal base plane. Horizontal machining centres are more common than the vertical machining centres in an FMS environment. They can be of three, four or five axes [4]. They are preferred for large, multi-sided workpieces as they have no restriction on the workpiece height and are useful for machining complex profiles. Horizontal machining centres with a vertical fourth axis can have multiple fixtures mounted on a tooling-block and can process several different workpieces in a single set-up.

Stiffness is an important attribute of a fixture body. The fixture body must limit deflection under the action of the cutting forces and clamps to maintain accuracy of the workpiece. It must also support the workpiece and provide sufficient system stiffness to avoid chatter vibration problems. Dedicated fixtures can be designed to satisfy the conditions for stiffness but with many flexible fixtures, lack of stiffness can present a problem.

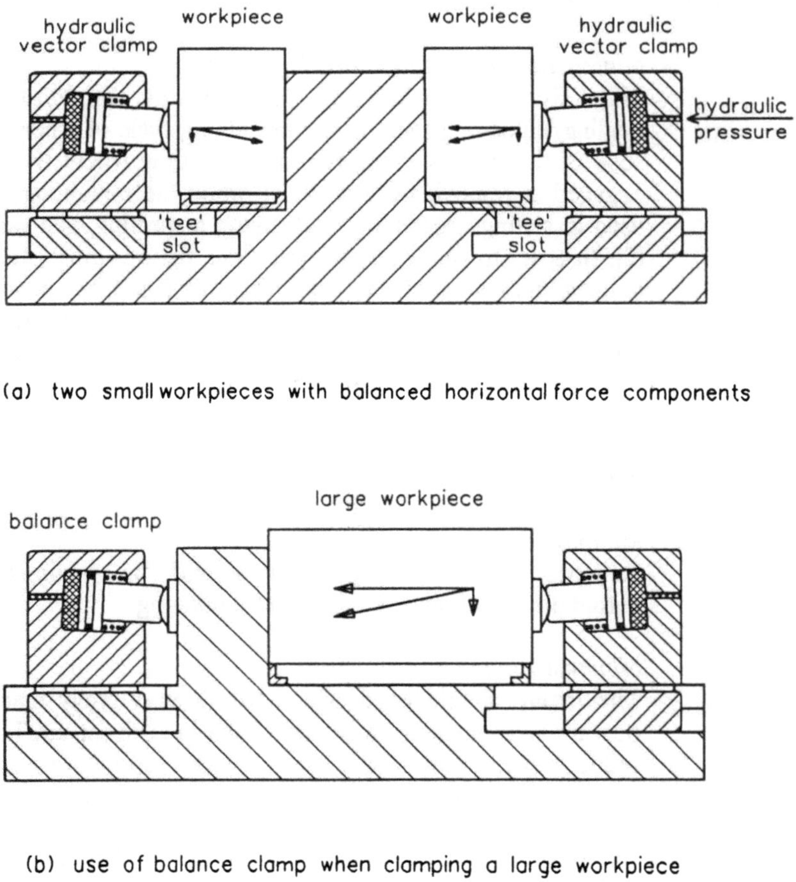

Figure 4.5 Vectored clamps

4.2.4 Short-Term Flexibility

To be viable in an FMS environment fixture systems must provide high short-term flexibility. Dedicated fixtures achieve this by means of off-line set-up of the fixture on pallets that can be quickly established on the FMS using AGVs and automatic pallet exchangers. This strategy can also be used for flexible fixtures but it results in duplication of fixture system elements and increased amortization factors. The ideal flexible fixture system should be capable of on-line reconfiguration (that is reconfiguration at the machine-tool), to accept any of the workpieces processed on the FMS.

Other fixture considerations related to short-term flexibility are:

- *Minimised part loading/unloading time:*
  The fixture should be designed to minimise part loading/unloading time. In the machining cycle of a part, about 30% of the time is spent as productive time where the tool is engaged in removing the material, while the remaining 70% is used in setting-up, loading, unloading, inspecting, idle time, *etc* [5]. In order to cut down the cycle time, fixtures must offer easy loading/unloading of the workpiece. For a CNC machining centre with two pallets changeover mode, loading/unloading of the parts can be accomplished while the system and machine tools are in cycle. Nevertheless, the fixture should be designed so that the workpiece can be easily and quickly loaded/unloaded to minimise part handling time, as in some cases the cycle time may be very short.

- *Minimized operator skill*
  Reliance on skilled operators should be minimized by automating workpiece loading and unloading.

- *Minimised number of set-ups*
  The fixture should be designed to allow as many machining operations as possible to be completed in each set-up. This will reduce part movements from one set-up to another. Consequently, errors due to change in location can be minimised. However, this may limit the area available for clamping. This constraint may compromise clamping stiffness and thereby reduce machining accuracies and metal removal rate. More machining passes with smaller depths of cut will be required to accomplish the machining dimensions which will increase the machining time.

- *Minimized build-up of error*
  Fixtures in an FMS environment may have to be transferred to and from processing stations via AGVs. The design should enable fixtures to be located and attached easily and accurately on to the machine table and the AGVs. Although fixtures can be positioned on the pallet which is designed and manufactured to be interchangeable, a cumulative error is possible between a fixture to a pallet and the pallet to a machining centre. Thus, depending upon the required tolerance, some fixtures must be fastened to a particular pallet throughout the operation to prevent the built-up error.

4.2.5 Long-Term Flexibility

Because their design can be optimised, specially designed, dedicated fixtures will invariably provide superior performance and short-term flexibility. To be viable flexible fixtures must be capable of accepting new part designs considerably faster than the lead time to produce a dedicated fixture and have a lower overall cost.

4.3 Flexible Fixturing

Current flexible fixture technology is summarised in Figure 4.6. Of these technologies, only modular fixtures and certain phase-change fixtures using authentic phase-change materials are mature and commercially available, the other technologies are the subject of research or are under development.

4.3.1 Modular Fixtures

Modular fixtures consist of a set of standard modular components or elements that provide the fixture functions of locators, clamps, supports, and the fixture body. These components can be assembled much like an 'adult Lego set' to build a fixture capable of handling a wide variety of part sizes and shapes. An example of a modular fixture assembly is shown in Figure 4.7 reproduced from [6]. The modular components are generally built over a baseplate which can be fixed onto a machine table. These fixtures consist of the baseplate, supporting, locating, clamping elements and fixture accessories. The selection of a modular element for a fixture depends on three major factors:

- Function of the modular element
- Geometry of the workpiece
- Size of the workpiece.

The basic concept of modular fixturing was first developed during World War II by John Warton. In the interim, modular fixtures have become very popular and became widely used with the implementation of CNC machines and subsequently in FMS. Hoffman [7] has published a comprehensive review of available modular fixture systems and lists twenty independent manufacturers of such fixture systems. Modular fixtures are popular because they offer most of the desirable features of dedicated fixtures with the advantage of relatively easy reconfigurability. Although the initial investment for a modular fixture kit is quite high, the capital investment can be amortized over a wide range of product parts and they are proven to be cost effective for low volume products. Some of the fundamental design criteria of modular fixtures are presented below and shown in Figure 4.8.

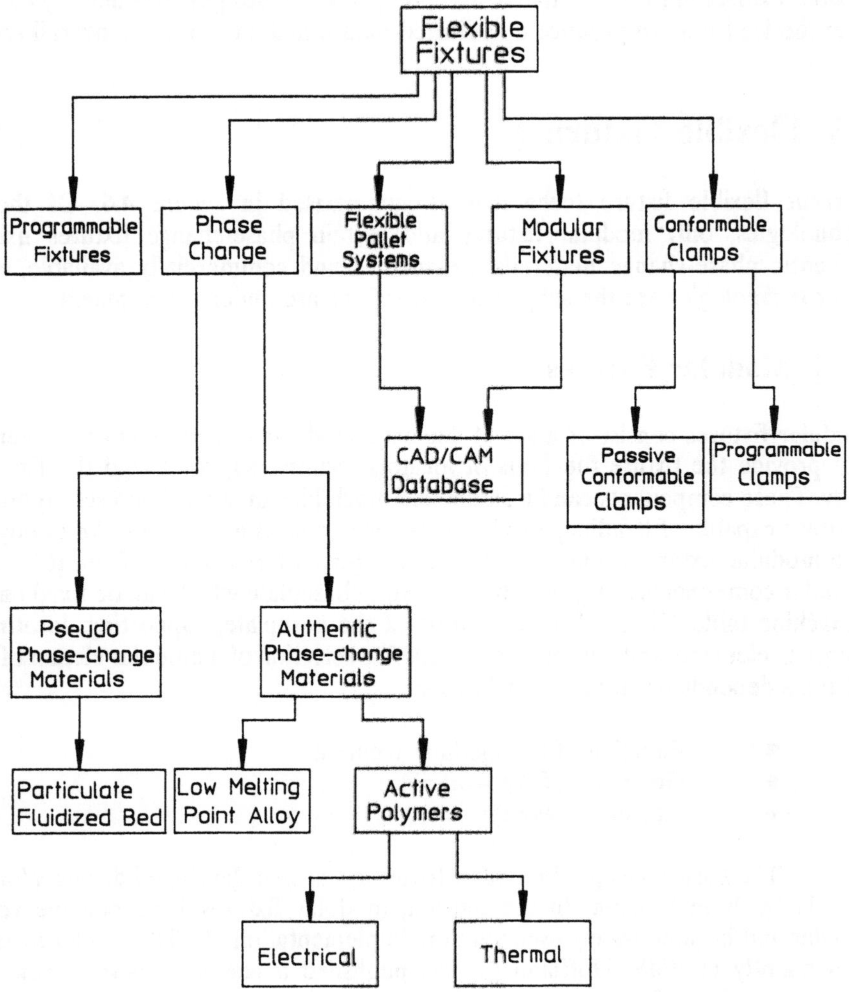

Figure 4.6 *An overview of flexible fixturing methodologies (Adopted from MV Gandhi and BS Thompson [16])*

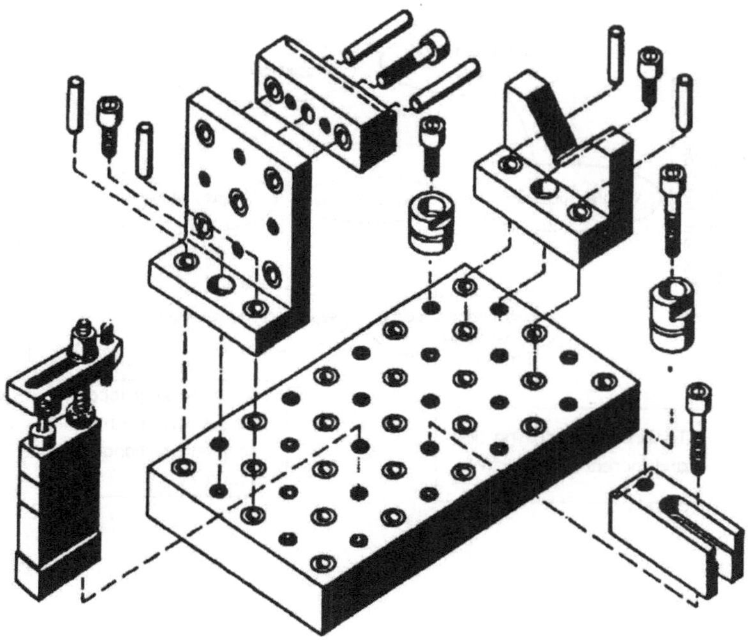

Figure 4.7 An example of a modular fixture assembly [6] (Courtesy of Blüco Technik)

- Positive location of the workpiece
- Rigidity of the workpiece
- Ruggedness of the fixture
- Repeatability of building a fixture (especially for hole-based systems)
- Versatility of the fixture
- Easy loading/unloading of the workpieces

Modular fixtures which satisfy the above fundamental design criteria are appropriate for the following applications (Figure 4.9):

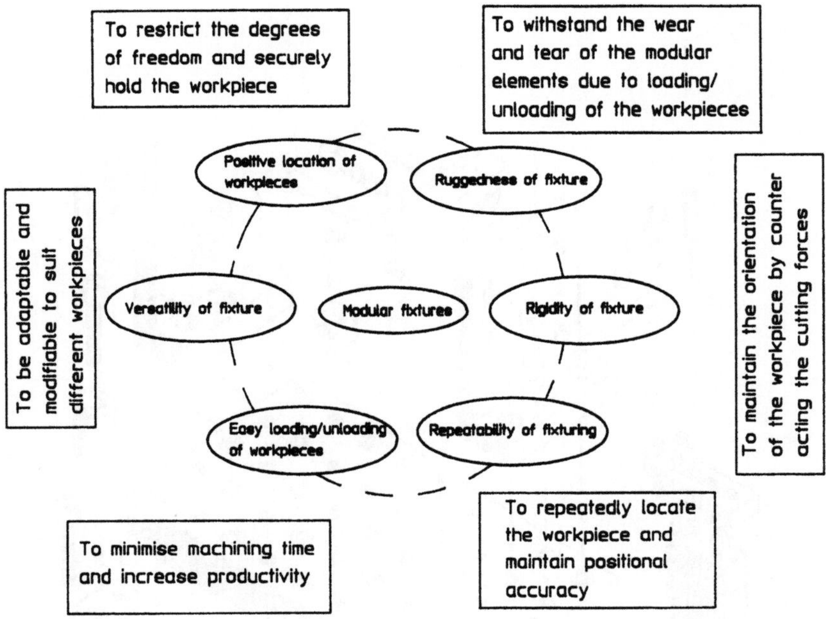

Figure 4.8 Fundamental design criteria of modular fixtures

- *Prototype tooling:*
  The modular fixturing system enables the tool designer to experiment with several ideas and designs before making a final decision. Tool design engineers can easily build up a fixture using modular fixtures to visualise their designs without incurring manufacturing costs.

- *Short run production:*
  Modular fixturing systems can maintain the necessary workpiece tolerances and accuracy with a fraction of the cost that is required for a dedicated fixture in the long run. In addition, time required for building a modular fixture is much shorter than a dedicated fixture.

- *Stand-by purpose:*
  Modular fixtures can be swiftly assembled to temporarily replace a dedicated fixture while it is being repaired, reworked or

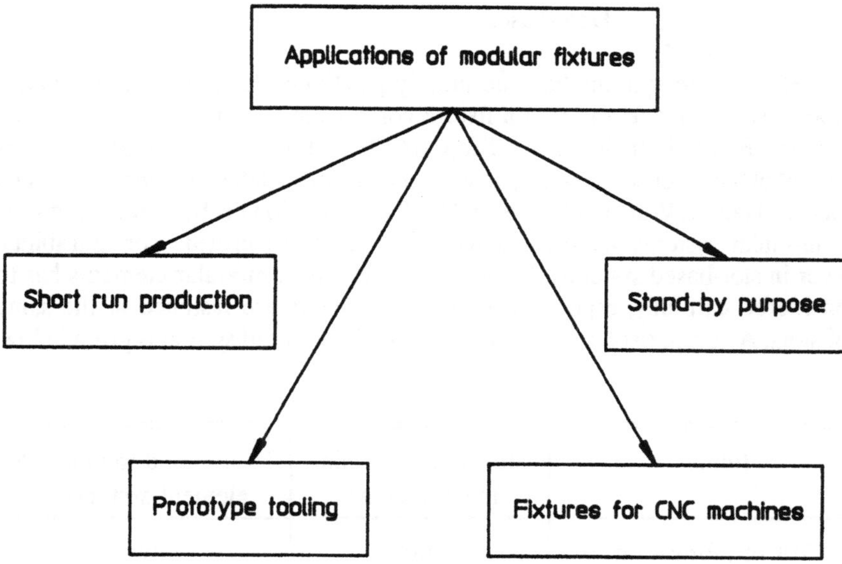

Figure 4.9 Appropriate functions of modular fixtures

revised.

- *Fixtures for CNC machines:*
  Modular fixturing offers the versatility to accommodate a variety of part families. Thus, this is very appropriate for CNC machine applications.

Some of the advantages of modular fixtures in an FMS include:

- Improving productivity by decreasing the lead time
- The ability to handle a variety of part families by increasing the adaptability
- Reducing the storage costs and improving efficiency by providing reusability
- The backup capability to reduce the breakdown time of fixtures

Commercially available modular fixturing systems can be classified into two types namely,

- Slot-based
- Hole-based

Hole-based systems have accurately positioned holes on the baseplates which are used to locate and fasten fixture components. Some examples of such systems are Blüco Technik, Imao, Kipp, SAFE and Venlic. Slot-based systems have parallel and perpendicular tee-slots on the baseplates. Examples of such systems are Halder, Warton Unitool and CATIC. Functionally both hole- and slot-based modular systems serve the same purpose *i.e.*, to provide configurability. However in slot-based systems, the order of assembly of modular elements has to be considered carefully, especially when the elements are fastened in the same row of slots. A comparison between the slot and hole-based system is given below.

Items	Slot-based modular element systems	Hole-based modular element systems
Cost of fabrication	High	Low
Ease of assembly and flexibility	Good	Relatively restricted
Requirement of skill and assembly	High	Relatively low
Adjustment of relative position of locating elements	Convenient and adjustment is not limited	Not convenient and adjustment is limited
Accessories for mounting the elements	More	Relatively few

A third type using a magnetic base which overcomes some of the constraints inherent in the above systems has been proposed by Asada and By [8]. This system has been specifically designed for assembly using robots and has as yet to be commercially exploited.

In general, the elements of a modular fixturing kit may be divided into three groups according to their function:

- baseplates, riser blocks and tooling blocks,
- locators, clamps and supports,

- tooling blocks, modular blocks and shims

Typically all the components of modular fixture systems are machined from stress-relieved meehanite, cast steel, or sometimes aluminium castings. The component parts are surface hardened to minimise wear, and surface ground to accurate size so that stack-ups of tolerances when the parts are assembled are kept to a minimum. They are usually machined with a flatness of 0.021 mm/m, parallelism of 0.042 mm/m and perpendicularity of 0.042 mm/m [4].

Functions of the modular fixture elements are briefly discussed below. A comprehensive practical description of commercially available systems is available in ref [7]

4.3.1.1 Baseplates, riser blocks and tooling blocks

The function of these elements is to provide an accurate surface for mounting the other fixture elements and an interface surface between the fixture and the machine-tool table or pallet. They provide the base structure for the fixture body. Examples of these elements are shown in Figures 4.10-4.13.

A baseplate for a slot-based system consists a rigid cast steel or meehanite plate with a number of tee-slots machined at right angles across the face of a baseplate as shown in Figure 4.10. Regardless of the baseplate shape, the tee-slots are machined perpendicular and parallel to each other to ensure accurate and precise alignment of the fixture elements. Circular plates used as face plates in turning centres have radial and tangential slots. The attachment of each fixture element is done by inserting a tee-clamping block into the slot and then firmly clamping the element in place with a high tensile socket headed bolt, or alternatively a high tensile bolt with a tee shaped head with a conventional nut is used. Most of the fixture elements in a tee-slot system can be keyed to the baseplate to ensure orthogonal alignment. When there are many fixture elements on the baseplate, interference between components can make it quite difficult to locate the tee-clamping blocks on the baseplate. An advantage of slot-based systems is that they provide infinite adjustment of mounting position in one direction, and also the chips generated during machining can be easily washed from the base plate.

Baseplates for hole-based systems have a matrix of precisely positioned holes as shown in Figure 4.11. The attachment of each fixture component is accomplished using socket head cap screws and inserting dowel pins in the pre-machined holes. Since this system utilises screws to build a fixture, re-arrangement of the components is easily achieved. This system allows only discrete adjustment of mounting position and unless more then one mounting hole is used to locate each element, orthogonality is not guaranteed. Chips and other

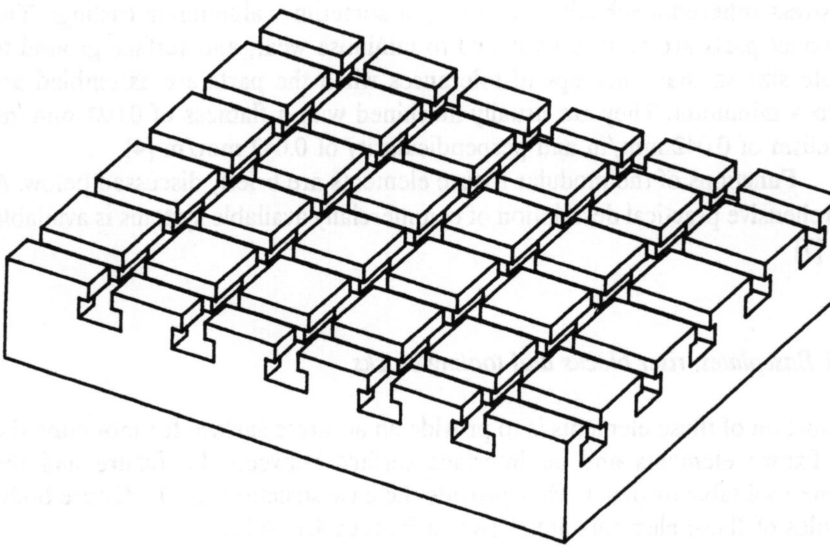

Figure 4.10 Baseplate of a slot-based system

machining debris tends to accumulate in the holes. This can be avoided by using a cap to plug unused holes.

Riser blocks are essentially extra thick baseplates. They are used to elevate the workpiece to eliminate the unusable dead space between the centre-line of the spindle and the top of the pallet or the machine table.

The main purpose of tooling blocks is to provide mounting surfaces at 90° to the base surface. They are frequently used to mount fixtures for small components on horizontal spindle machining centres. Different fixtures can be mounted on each surface of the tooling block. A two-sided tooling block is shown in Figure 4.12. When such a tooling block is mounted on a rotary table, it can be indexed 180° to present two work set-ups to the cutting tools in rapid succession. The four-sided tooling block in Figure 4.13 can present four surfaces to the spindle.

Tooling blocks, because of their high cost of manufacture, appear to be limited to the hole system. However tee-slot baseplates can be attached to their vertical surfaces.

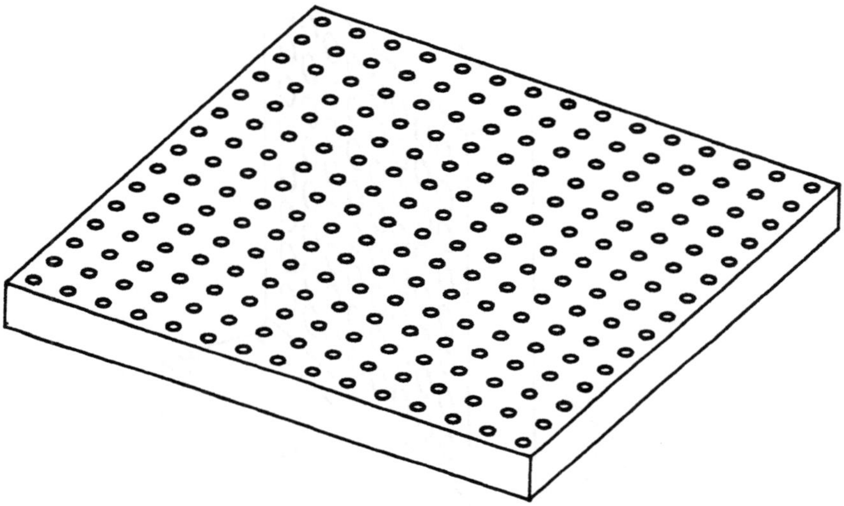

Figure 4.11 Baseplate of a hole-based system

4.3.1.2 Locators supports and clamps

Locators supports and clamps used in modular fixtures are essentially similar to those described in Chapter 2 in relation to dedicated fixtures and shown in Figures 2.3, 2.4 and 2.5.

In hole-based modular fixture systems, these components are usually mounted directly to the horizontal surface of the baseplate or a tooling block. Locators and supports are of a different design depending on whether the surface to be located (or supported) is parallel or perpendicular to the mounting surface. The common locating and supporting elements of a standard hole-based system are shown in Figure 4.14 and include:

- Surface edge bars and blocks as shown in Figure 4.14 (a,b). These are used as risers and edge locators either individually or in combination with other components of the system having the same hole pattern. Vertical edges and horizontal faces can be located simultaneously.

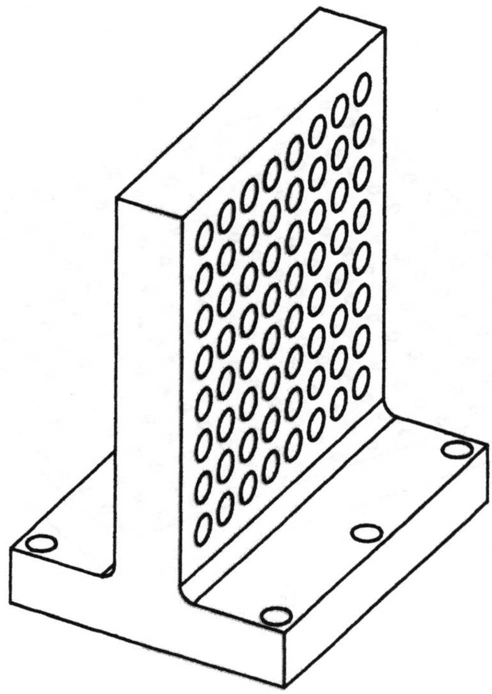

Figure 4.12 Two-sided tooling block of a hole-based system

- Adjustable surface locators are useful as location and support points as many of the workpieces to be fixtured may not fall into the exact hole-matrix pattern. Some locators have additional functional features such as a ball at the locating end as shown in Figure 4.14 (c). This will help the locator to locate irregular surfaces such as castings and forgings.

- V-locators, as shown in Figure 4.14 (d), are used to locate curved surfaces. Different types of V-locators include 120° V-blocks and 150° V-blocks. A 120° V-block can be used to locate surfaces up to 200mm diameter. However larger diameters can be handled with 150° V-blocks.

- Some locators have a flat face, as shown in Figure 4.14 (e) to locate vertical edge surfaces.

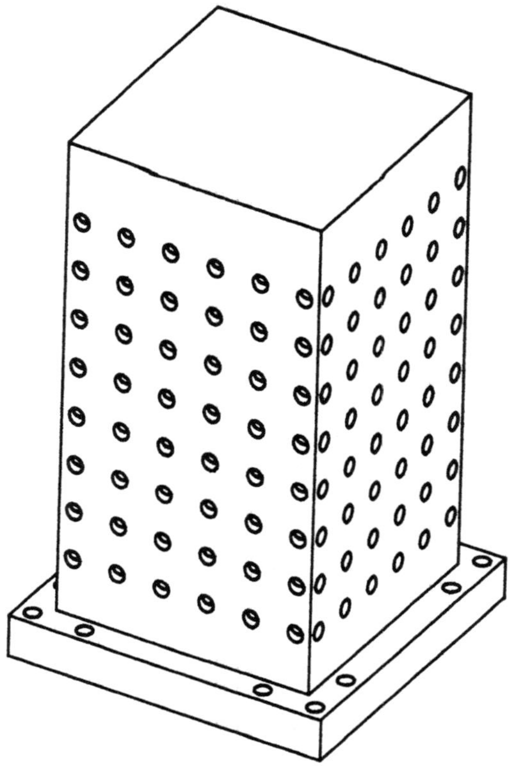

Figure 4.13 Four-sided tooling block of a hole-based system

- Additional locating pins, as shown in Figure 4.14 (f), are available to support or locate horizontal face surfaces.

Tee-slot systems may use similarly designed locators, supports and clamps but there is more general reliance on round or flat button locators for locating both vertical edges and horizontal faces. This is made possible by using towers of modular blocks as described in the following section and illustrated in Figure 4.15. Button locators have the advantage of more exact conformance with the ideal 3-2-1 system of location. By using shims they can be set accurately in the first place and adjusted to correct subsequent wear.

For design purposes, locators, supports, and clamps may be categorized as follows, according to their use.

- horizontal face locators, clamps and supports

  These attached to the body of the fixture at a horizontal plane and are generally subjected to vertical forces only.

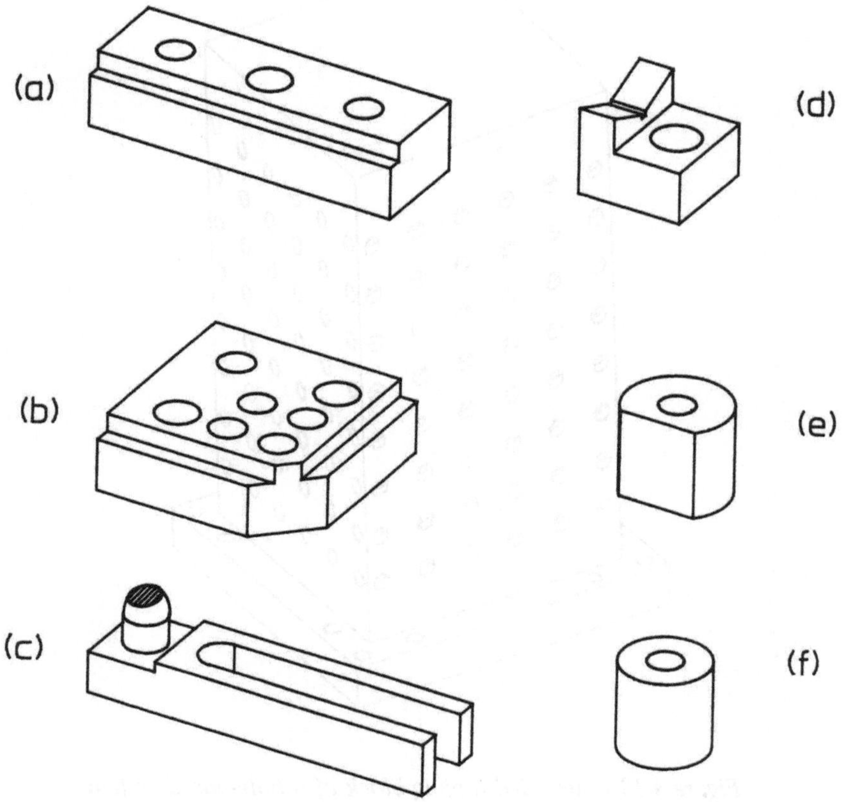

Figure 4.14 Common locating and supporting elements of a hole-based system

- vertical face locators, clamps and supports

  These attached to the body of the fixture at a vertical plane and are generally subjected to horizontal forces only

- thrust locators

  These are a special case of the vertical face locator. They are placed in line with the highest machining or clamping forces and for maximum stiffness they must be attached to the baseplate in as direct a way as possible

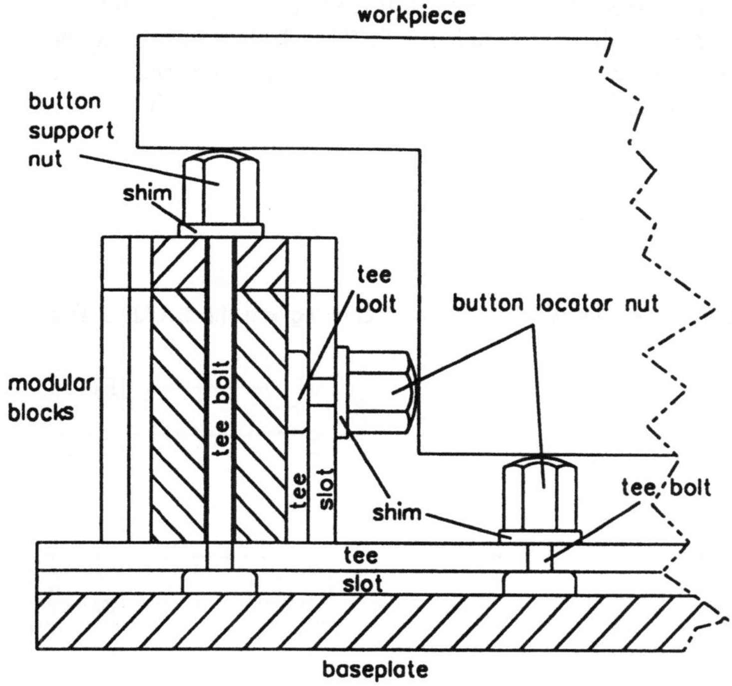

Figure 4.15 *Use of button nuts as locators and supports in a 'tee'-slot modular system*

Most of the standard types of clamping elements used in dedicated fixtures are available in modular fixturing systems. They include the hydraulically activated clamps, adjustable vice jaws, bridge clamps (top clamps), strap clamps, screw clamps, toggle action clamps , *etc.* The type of clamping element used depends on the geometry of the workpiece and the forces involved. Modular systems have the advantage that they are produced in large numbers, and automatic and power-operated clamps that would be prohibitively expensive for a dedicated one-off designed fixture can be part of a standard kit of parts. Such standard clamping systems often incorporate design improvements that would be difficult to realise in a special purpose fixture. Examples of this are the vectored clamps discussed previously and the De-Sta-Co SAFE self adapting clamps shown in Figure 4.16. This system uses ball elements on its clamping surfaces and has the ability to automatically adjust to accommodate imperfect workpieces such as castings or forgings, which can often be warped. If conventional strap clamps are used, the warped surfaces may lead to uneven clamping at the support and clamping points as shown in Figure 4.16 (a). When clamping forces are applied, severe strains are produced as shown in Figure 4.16 (b). When the clamps are released the machined workpiece will spring back as in Figure 4.16 (c), resulting in rework. Distortion due to clamping is further elaborated in Chapter 6. The SAFE clamping system does not distort the workpiece and the surface of the workpiece is machined perfectly flat as shown in Figure 4.16 (f).

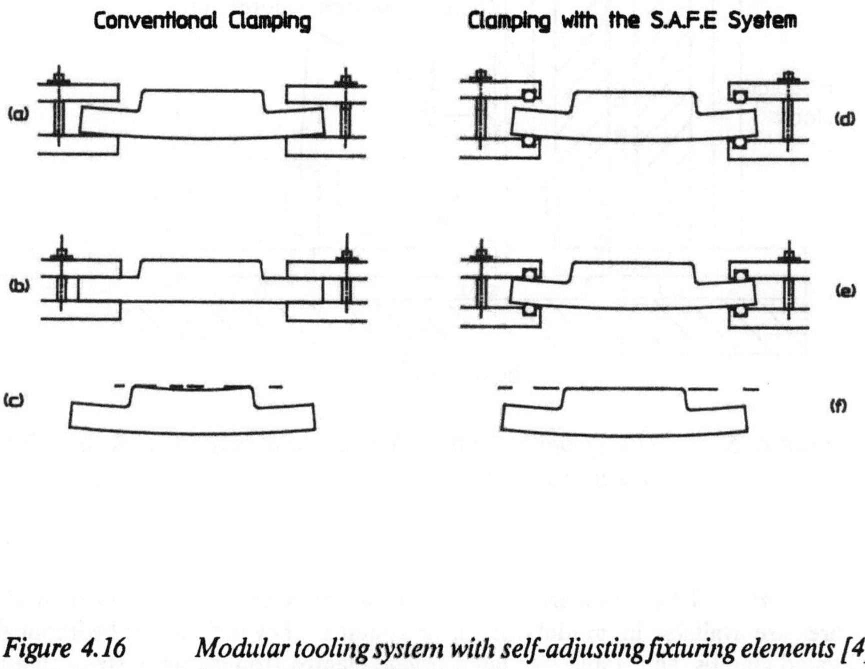

Figure 4.16 *Modular tooling system with self-adjusting fixturing elements [4] (courtesy of S.A.F.E system)*

4.3.1.3 Tooling cubes, modular blocks and shims

Tooling cubes are similar to small tooling blocks. They are used in hole-based modular fixture systems to provide vertical surfaces for mounting locators, clamps or supports. They are usually mounted directly to the baseplate or a pallet. An example tooling cube is shown in Figure 4.17. The tooling cube is hollow to reduce the overall weight while remaining rigid enough to counteract the cutting forces during machining.

Much of the design flexibility of slot-based modular systems comes from the use of modular blocks that are used to form the fixture body which provides

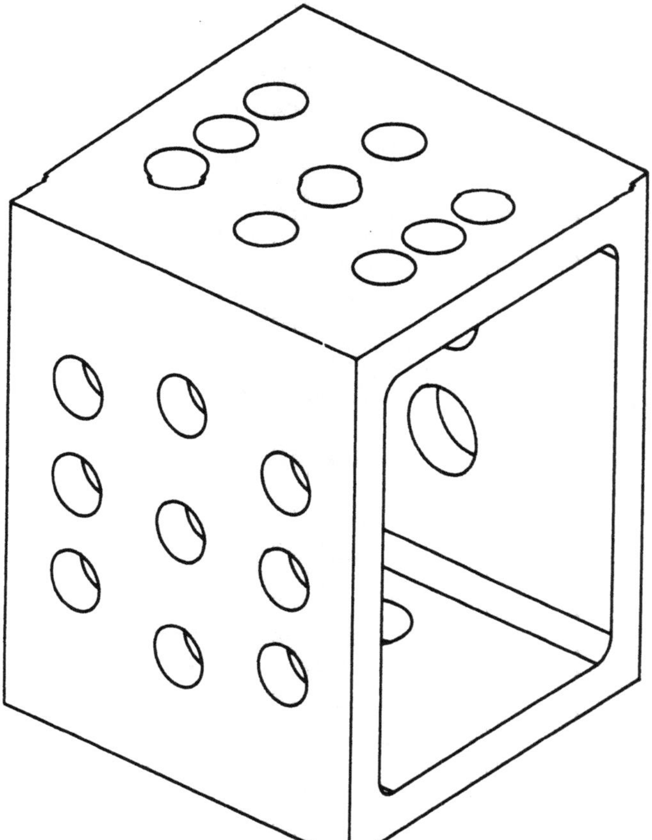

Figure 4.17 An example tooling cube

a connection between the baseplate and the locators, clamps and supports. The modular blocks used in the CATIC system, for example, are of 60mm square cross section and of various heights. Each block has a vertical hole passing through the 'top' and the 'bottom' faces. The other four vertical faces are machined with slots parallel to the axis of the hole. Figure 4.18 shows a schematic drawing of a typical modular block. The topology of the blocks has important implications for connectivity.

In the CATIC system, the slot on the modular block has restricted connectivity because it can only be connected to a hole. However it allows sliding (continuous movement) along the slot. In contrast with the slot, the hole in the modular block provides more flexible connectivity because almost any face can be connected to a hole. However, it is inflexible in terms of relative movements between modular elements. Different configurations are used in different modular fixturing systems. The Halder system allows slot to slot connection using clamping blocks as shown in figure 4.19, but the basic concept of the slot and holes is the same. Knowledge of constraints of this type must be incorporated in any design system for the fixture body.

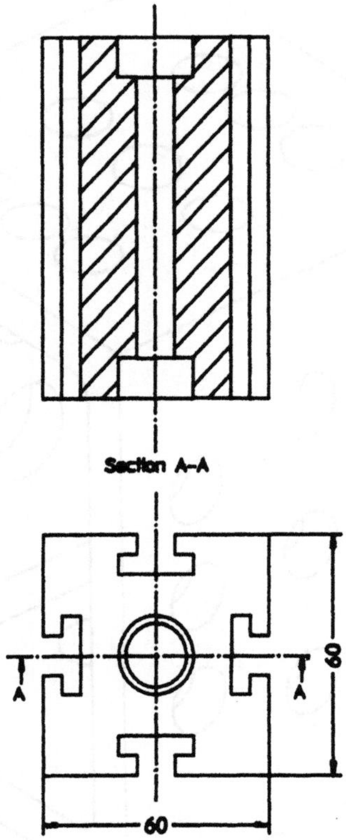

Figure 4.18 A typical 'tee'-slot modular block

Combinations of blocks are used to provide continuous freedom in placing other elements anywhere in the x-y-z workspace. Keys are placed in the slots to maintain orthogonality throughout the structure. There are several fundamental structures of combinations of blocks used to provide the required degrees of freedom.

The three-block structure, as shown in Figure 4.20, is the most general structure. Each block may consist of a few modular blocks and shims held together by a single high tensile bolt. The combination of modular blocks and shims gives the required length of a block. The three blocks in orthogonal arrangement are able to reach any point on the surface of the workpiece from a slot of the baseplate. The 'x-y-z' dimension is achieved by adjusting the length of the blocks in two directions and sliding the structure along the tee-slot of the baseplate in the third direction. The connectivity constraint of the structure is satisfied by the orthogonal arrangement.

The three-block structure consists of a tower and a mounting block. The block that is attached to the baseplate is the mounting block while the other two

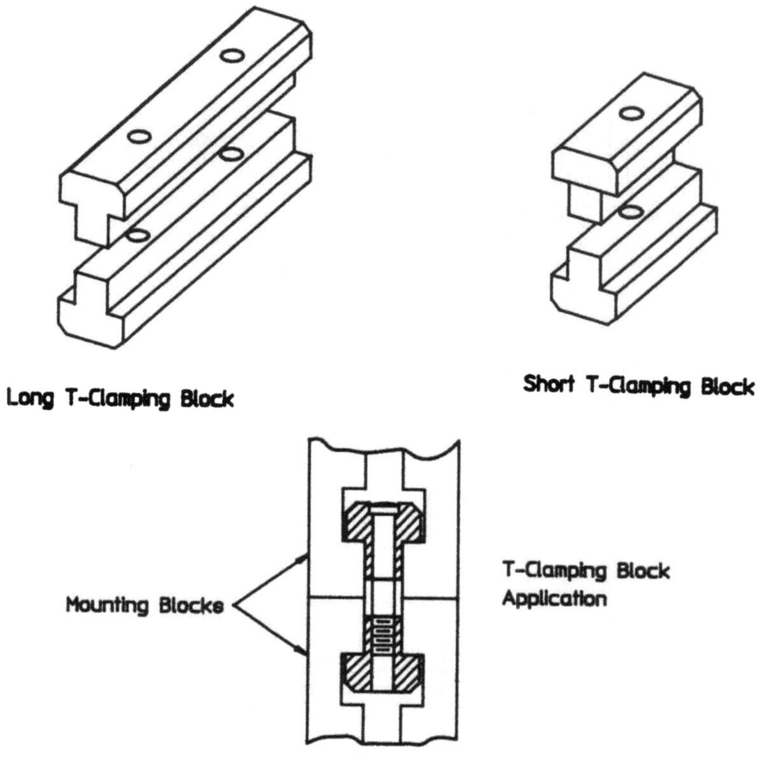

Figure 4.19 Tee-clamping block used for slot-to-slot connection

blocks form the tower. This tower may hold either a vertical face or a horizontal face locator, clamp or support.

Besides the general three-block configuration, in some cases two-block structures or even one-block structures may be used. The two-block structure is shown in Figure 4.21. This structure provides two degrees of freedom. The mounting block can slide along the slot in the baseplate and fixture elements can be clamped anywhere along the slot in the second block. This structure occupies less space but can only be used if a horizontal face locator, clamp, or support is positioned at a height above the baseplate equal to the width of a block.

In a one-block structure, the single block serves as both the tower and the mounting block. The advantages of using a one-block structure are:

- It does not require a separate mounting block.
- It is connected directly onto the baseplate allowing maximum rigidity.

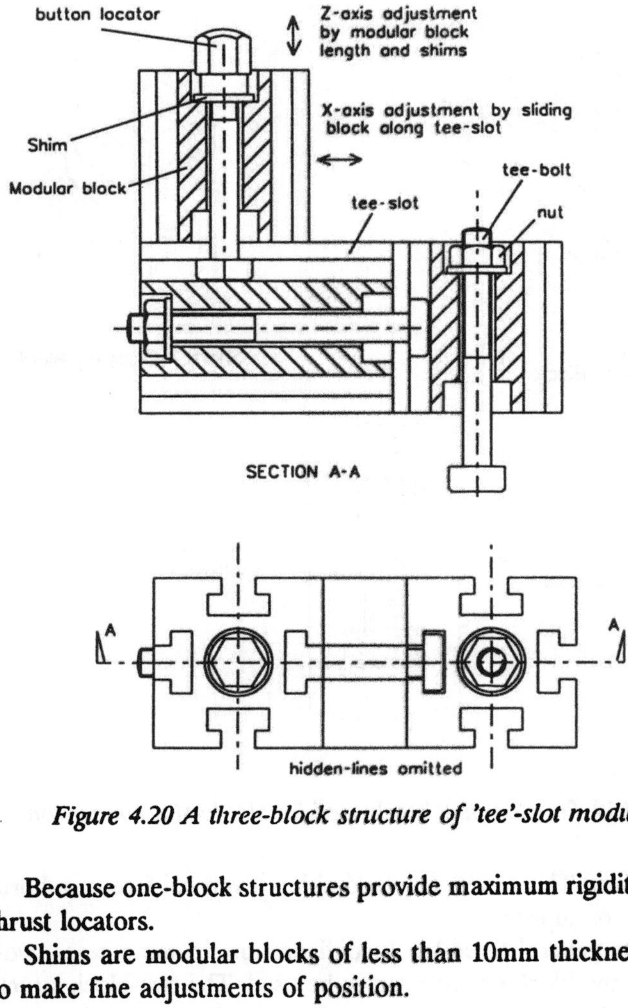

Figure 4.20 A three-block structure of 'tee'-slot modular blocks

Because one-block structures provide maximum rigidity, they are used to hold thrust locators.

Shims are modular blocks of less than 10mm thickness or thin washers used to make fine adjustments of position.

4.3.1.4 Evaluation of modular fixture systems

Modular fixturing provides a number of advantages, such as the reuse of fixture elements and low amortization factors. Hoffman [7] has reported that the capital cost of fixtures constructed from modular elements is approximately 25% of an equivalent dedicated fixture. They enable easy modification of the fixture, reduction of storage space and cost incurred while the fixtures are not in use. There is no doubt that they allow considerable improvement in system reconfigurability with very little loss of performance when compared with dedicated special purpose fixtures. With the present state of development of alternative flexible fixture systems, modular fixtures offer the most promise for

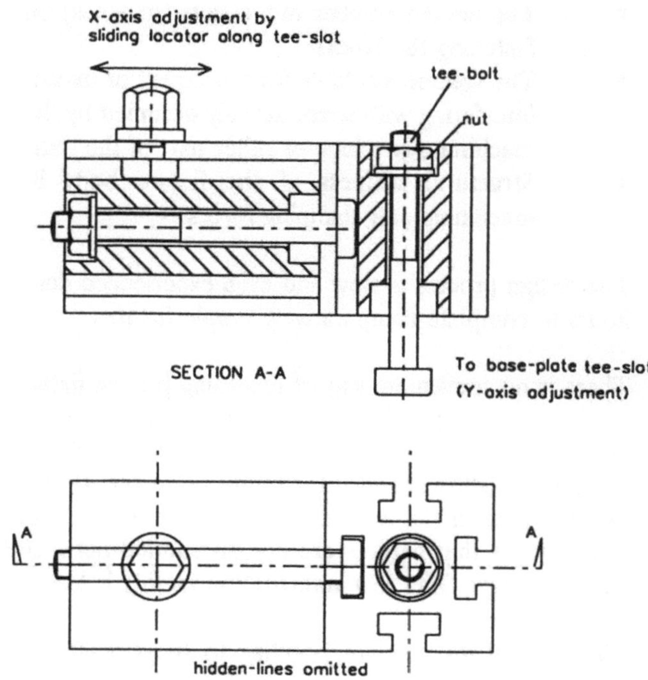

Figure 4.21 A two-block structure of 'tee'-slot modular blocks

increased reconfigurability of FMS.

There are however some problems. A postal survey of U.K. users of the CATIC modular fixture system, conducted by Whybrew and Ngoi [9], revealed the following problems:

- The design of modular fixtures lacks systematic procedure. It is a general practice for the fixture design and construction to be performed simultaneously in the toolroom by experienced operators. The procedure is based on experience and 'trial and error'.

- Design of towers of modular blocks is difficult. The process planner will usually indicate location, support and clamping points on the workpiece and it is the designer's task to select appropriate fixture elements and design towers of modular blocks to connect these elements with a baseplate. This is a formidable task and the designer must take into account:

  - The connectivity between modular blocks
  - The limited number of tee-slots (or holes) on the baseplate for fastening the blocks
  - The volume available for placement of fixture elements without interfering with space already occupied by the workpiece or the machining envelope or other part of the fixture
  - Structural stiffness of the fixture body in relation to the machining and clamping forces.

- The design process is slow and even experienced designers take several hours to complete comparatively simple fixtures.

- There is no convenient way of recording precise fixture assembly details for future reference. The 'trial and error' approach means that there are no drawings and usually only a list of parts and a photograph of the completed fixture are filed for future reference. If the fixture is stripped between batches this lack of information compromises repeatability. This often means that fixtures are not dismantled between batches and any cost advantage from low amortization factors is lost.

These problems are not peculiar to the CATIC system and similar problems have been reported with other systems [10],[11] and [12].

Computer aided design systems that are combined with knowledge based systems that have rules concerning connectivity and tower construction have been demonstrated to overcome these problems. Typical systems are by Farkus et al [13] and Ngoi [9], [14]. A typical design produced by Ngoi's program is shown in Figure 4.22.

Nee et al have developed a parts coding and classification system especially adapted to fixture design requirements [15]. Using this system, when machining a new part in a family of parts, the workpiece code can be used to retrieve the fixture details of a similar existing design. Modifications made to an existing design are generally much faster than a completely new configuration of the modular fixture system. This method is discussed in detail in Chapter 5

Another problem with commercially available modular fixture systems is that they are designed for manual assembly and are too complex for robotic assembly. A variation of the hole-based system using special serrated washers has been developed by Woodwark and Graham at the University of Bath [10]. This system is designed for ease of handling and assembly. The fixture shown in Figure 4.23 is constructed in a 'pancake' fashion from vertical stacks of washers. Orientation is maintained by means of interlocking serrations on the mating faces. The top of each stack is a locating or clamping element.

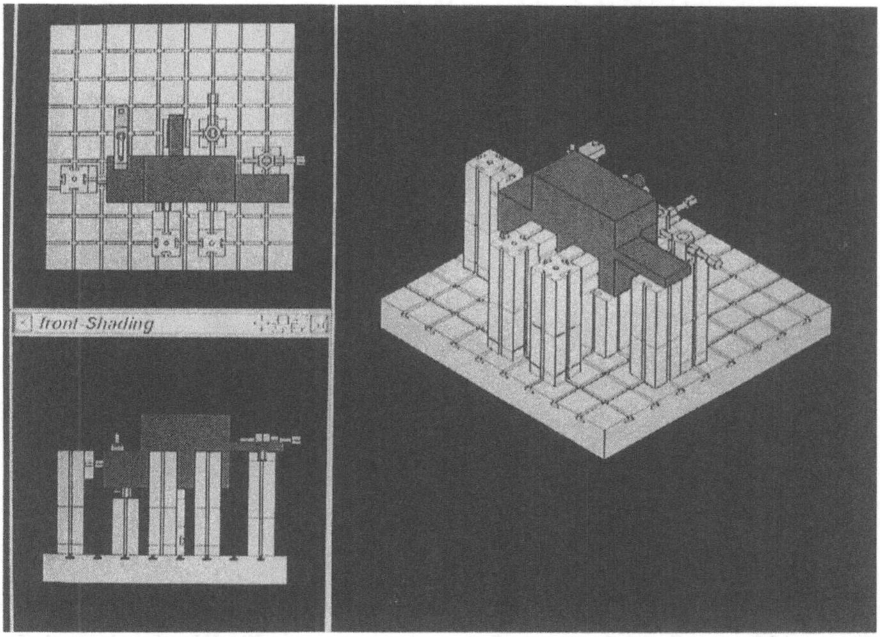

Figure 4.22 *Computer generated assembly of modular elements (courtesy of Ngoi [9])*

4.3.2 Flexible Pallet Systems

Flexible pallet systems are a variation of the modular fixture approach. Typical systems are by AIOI Engineering Inc.[3] and The Mors Group Inc. [16]. These systems consist of a range of standard pallets to which can be attached a range of standard fixed location blocks and repositionable clamps. They are usually used to hold several workpieces as shown in Figure 4.24. These systems can incorporate a high level of automation. In the Mors system, for instance, the pallet has a self-contained, programmable hydraulic clamping system. Each pallet has an on-board microprocessor to control the clamping cycle and pressures. Data and clamping

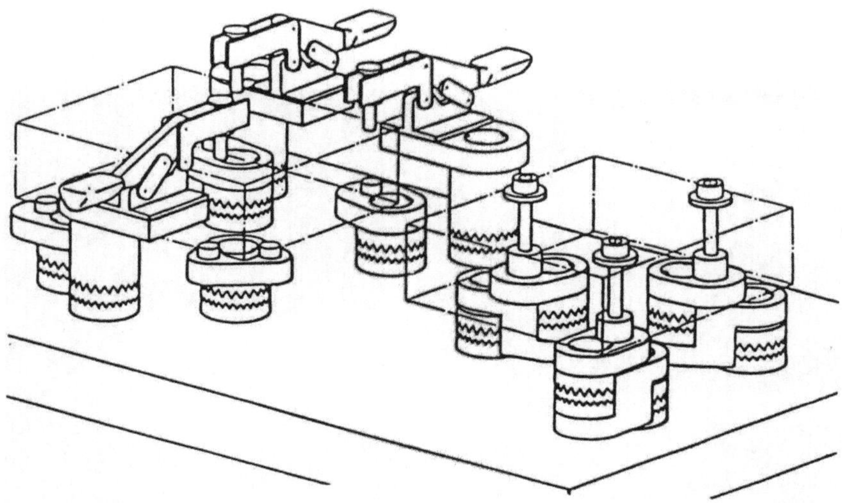

Figure 4.23 *Special-washer fixture (courtesy of JR Woodwark and D Graham [10])*

programs are transmitted to the pallets by an infra-red beam.

4.3.3 Phase-Change Fixtures

Phase-change fixtures use the concept of a material phase change [16]. This technique makes use of certain class of materials such as a low melting point alloy which is capable of rapidly converting its phase from liquid to solid and vice versa. Typically, a fixture of this kind consists of a container filled with this material and a mechanism to initiate the phase change. The fixturing procedure is initiated when the bi-phase material is in the liquid or semi-liquid state. A workpiece is immersed and placed in a desirable orientation in the container. The material is then subjected to some external influence (catalysts or cooling) which solidifies the material and firmly secure the part in the desired position for the machining operations. When the machining operations are completed, the material is once

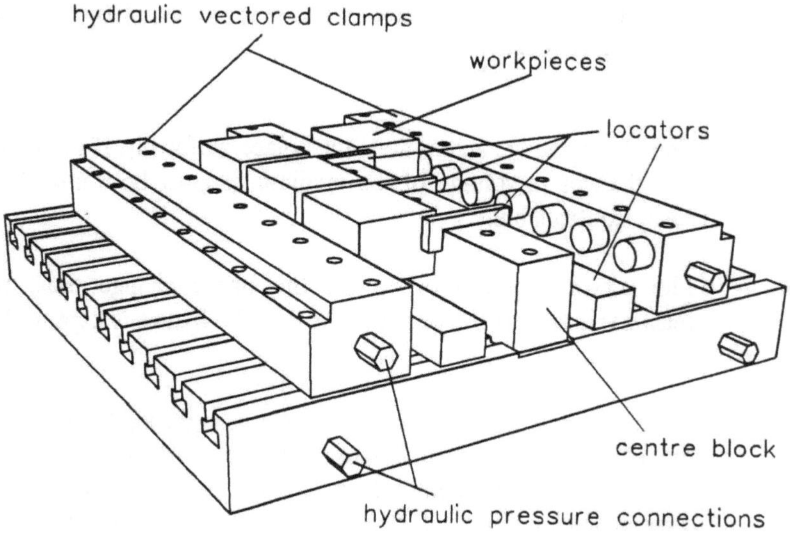

Figure 4.24 A flexible pallet system used to hold multiple workpieces

again subjected to catalyst actions to return to its liquid form and the part is easily removed from the fixture.

This type of flexible fixturing is appropriate for irregular workpieces which are very difficult to hold. However, the phase-change fixtures still have some disadvantages. These fixtures provide support but not location and some additional mechanism is needed to align the workpiece when it is immersed in the liquid medium. The location function is therefore transferred to a separate alignment jig or a robot to hold the workpiece until the material becomes solid and strong enough to hold and maintain the position of the workpiece. The cost and difficulty of reconfiguring the separate location mechanism remains.

Some of the evaluation criteria used in the selection of a suitable phase-change media are discussed below:

- The phase-change must be affected quickly, as slow phase-change will affect productivity.

- Since the material in the phase-change fixture must repeatedly change its phase, the material state must be consistently reversible over a large number of cycles without losing its properties or changing density. Otherwise, the processes are unlikely to be economical.

- The medium must not have an adverse effect on the surface finish or properties of the workpiece. It must not be toxic or detrimental to health.

- The phase-change action must take place uniformly. If there are any changes in volume of the medium, it must be minimal or easily controllable.

- The power required to initiate the phase change must be reasonable so that the operating cost of the fixture is kept to the minimum.

- If the temperature of the medium is changed during the process, the effect of the thermal stress on the workpiece must be analyzed as the temperature change is likely to affect dimensional stability of the workpiece causing warping and distortion.

Depending upon the characteristics of the process, the phase-change fixtures can be classified into authentic and pseudo phase-change fixtures

4.3.3.1 Authentic phase-change fixtures

Authentic phase-change fixtures refer to the use of a medium that physically changes its phase from liquid to solid and vice versa. The process of changing its phase can be activated by means of temperature or electrical field or a combination of both. For example, low melting point bismuth alloys (Cerrobend®) are currently used as an encapsulating medium in temperature induced phase-change fixtures. In electrically induced phase-change fixturing, modern polymeric materials such as polyacrylonitrile are used. The electrostatic forces due to the electrical field affect the intermolecular structure of the polymeric network thereby causing a phase-change. In some cases, an auxiliary temperature field can be used to accelerate the phase-change process.

- Encapsulation systems

Most commercially available systems work by direct encapsulation of the system. The method employed in the Fisher Fixturblok system is a typical procedure [17]. It has been developed specifically for machining the roots of turbine and compressor blades. The sequence of operations is illustrated in Figure 4.25. The blade is contained in a diecasting mould and precisely located with respect to the mould by an external alignment jig. Liquid low melting point alloy is injected into the mould and allowed to cool and solidify. The blade encapsulated in the block is removed from the mould. The encapsulation block has standardised location features so that it can be held in a standard fixture on the machine-tool. When machining is complete the block is cracked open to release the blade and the block material is re-cycled. The system works very well in this application which would be very difficult to fixture in any other way. There are drawbacks to the system. The external alignment jigs are a restricting factor for reconfigurability and the encapsulation material is not as stiff as conventional materials used in fixture construction. The process can only be used for small workpieces, 300mm or less. The mass of the block material is a limiting factor if robot handling is required.

- phase-change baseplate

This type of fixture, suggested by Ngoi [9] is illustrated in Figure 4.26. It uses many standard modular fixture components but in place of the usual baseplate with holes and slots the fixture elements are held in place in a rigid container by surrounding them with phase-changeable material. This system has the advantage of freedom in positioning elements without the constraint of holes or slot position. Towers of blocks can be simpler and more compact, but there is a fundamental problem of much lower stiffness when compared with fixtures built on solid baseplates. This system is not commercially available.

4.3.3.2 Pseudo phase-change fixtures

Pseudo phase-change fixtures utilise the two-phase nature of a particulate fluidised bed. The basic principle is suggested when handling vacuum packed bags of coffee. Before the vacuum seal is broken these packs are solid and quite inflexible, but when the seal is broken and air enters the bag and the bag becomes very flexible.

- Elastic robot grippers

This mechanism has been investigated by Perovskii [18] as a robot gripper for irregularly shaped workpieces but it would seem also to have application as a fixture system for irregularly shaped workpieces in low-force machining applications. The basic structure of Perovskii's gripper is a flexible rubber bladder filled with sand, surrounded by a 'U' shaped

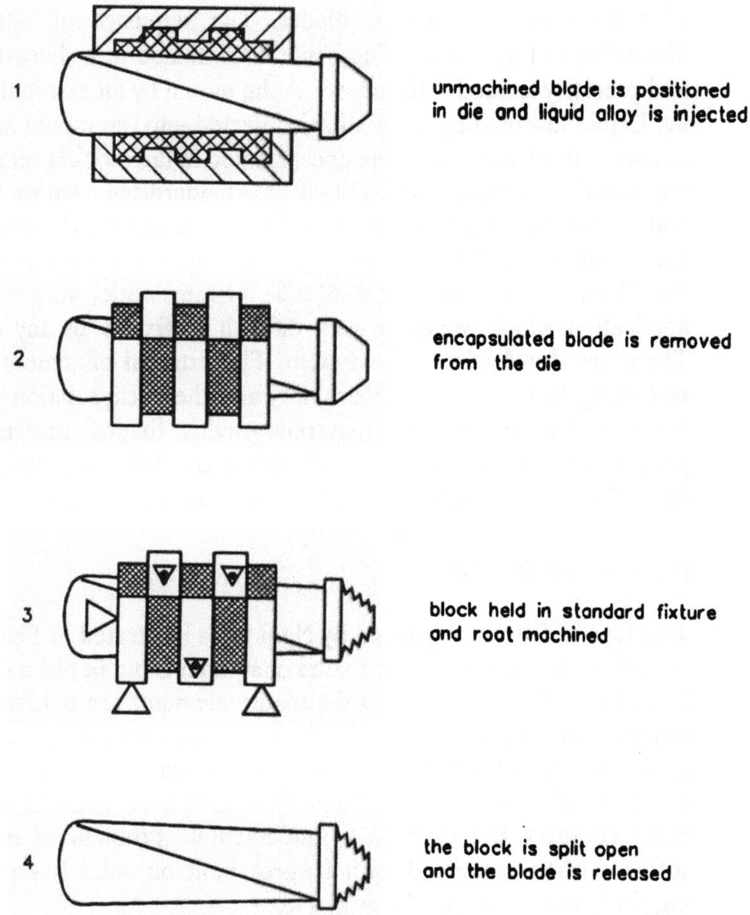

Figure 4.25 Encapsulation phase-change fixtures

semi-corrugated rubber hose as shown in Figure 4.27. The workpiece is placed in the open jaws of the gripper and compressed air is admitted to the outer semi-corrugated hose causing the external sides of the hose to expand. This forces the sand-filled inner bag to conform to the shape of the workpiece. The inner bag is then connected to a vacuum pump to evacuate the air from the bag. The grains of sand lock with each other to form a rigid structure. The workpiece is released by releasing the air pressure from the outer-bag and refluidizing the sand by re-admitting air to the inner bladder. This type of fixture in common with other encapsulating phase-change fixtures, provides support but no location.

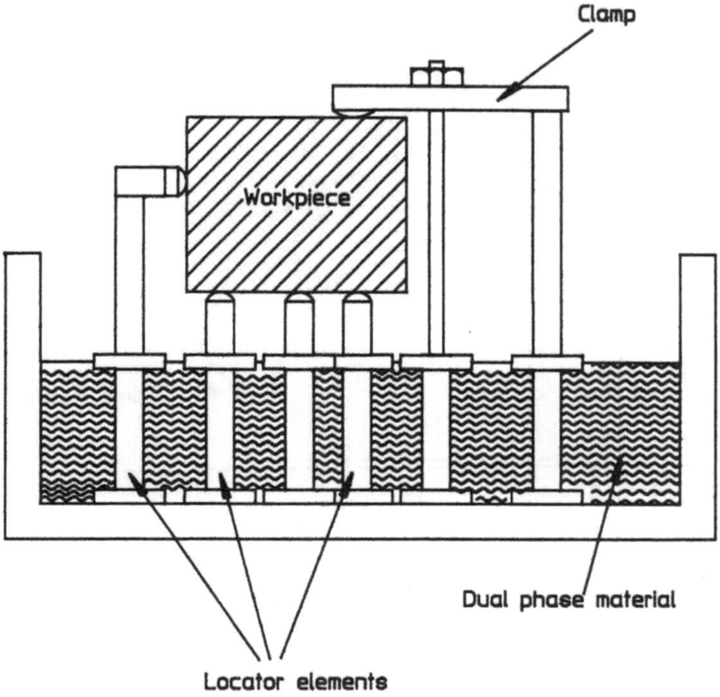

Figure 4.26 Phase-change baseplate concept (courtesy of Ngoi[9])

- Fluidized bed fixtures

A schematic arrangement of a fluidised bed fixture is shown in Figure 4.28. The fixture consists of a container filled with small particles over a porous floor through which an air stream passes at a carefully controlled rate. When the air stream passes through the porous floor, the particles are suspended in the air stream and therefore become quite loose. A workpiece can then be inserted into the container without much difficulty. When the air supply is turned off, the particles compact under gravitational force to form a solid mass which is capable of holding the workpiece securely. The particles can be further compacted by applying a normal load to the compaction plate which in turn increases the force induced on the workpiece. Once the workpiece is secured, the machining operations can be performed. After the machining operation, the workpiece can be removed by releasing the compaction load and fluidizing the medium. Because the process does not involve an actual phase-change from fluid to solid and vice versa, this type of fixturing

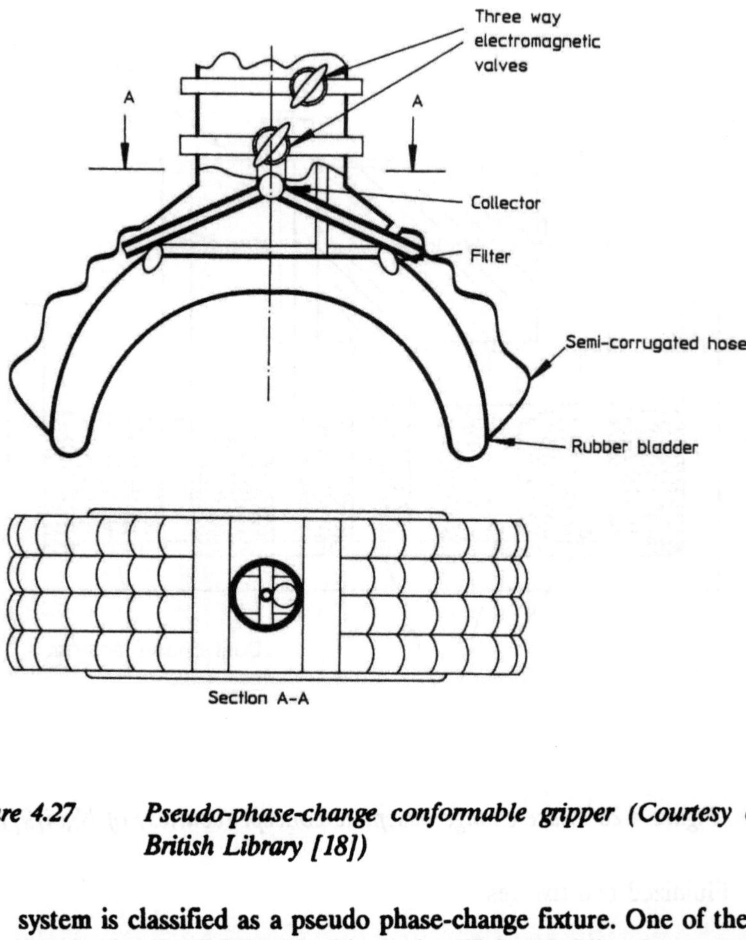

Figure 4.27 *Pseudo-phase-change conformable gripper (Courtesy of The British Library [18])*

system is classified as a pseudo phase-change fixture. One of the major drawbacks of a fluidized bed is the insufficient holding forces generated for most metal cutting operations [19].

It is obvious that the performance of a phase-change fixture in terms of the amount of holding force depends on several factors such as the height of the fluidised bed, and immersion length of the workpiece in the particles. Prototypes based on this concept are currently being developed. Several experimental and theoretical studies were carried out [16] to study the performance of fluidized bed phase-change fixtures.

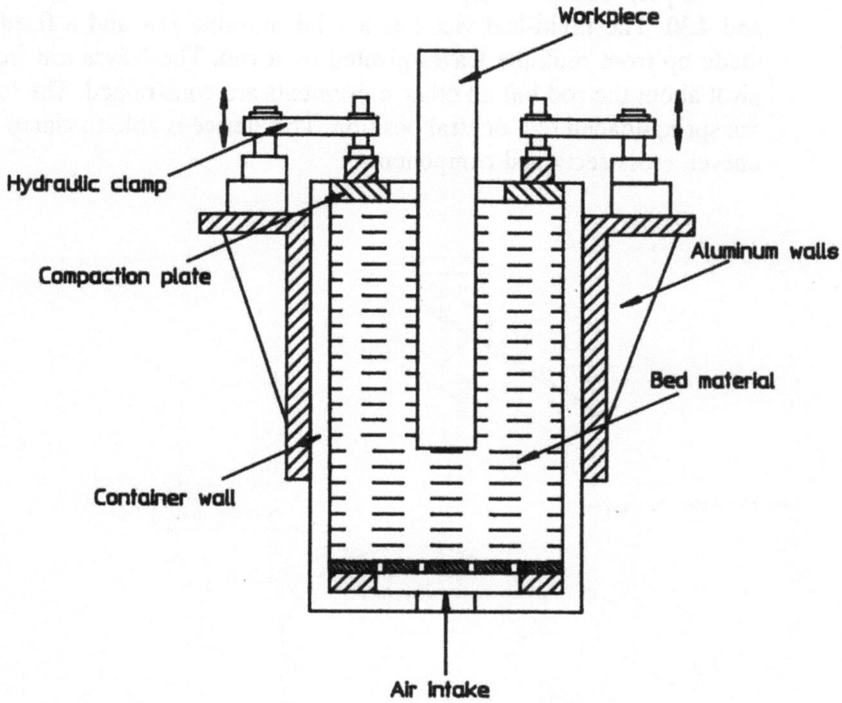

Figure 4.28 *Schematic of particulate fluidized bed fixture [4] (Reprinted with permission of the Society of Manufacturing Engineers, copyright 1989)*

4.3.4 Conformable Fixtures

Conformable fixtures are fixtures with clamping elements that automatically conform to the shape of the workpiece. They are passive device, which can change shape and reach a stable configuration when the clamping force is applied or they may be programmable to change shape under active control.

- Passive conformable clamps

B.S. Thompson has described two passive conformable fixtures in his review paper on flexible fixtures [1]. These are the multi-leaf vice and the petal collet. Thompson's drawings are reproduced in Figures 4.29 and 4.30. The multi-leaf vice has a solid movable jaw and a fixed jaw made up from multiple leaves pivoted on a rod. The leaves are free to pivot about the rod but all other movements are constrained. The leaves are spring-loaded to a neutral position. This device is able to clamp long uneven cross-sectioned components.

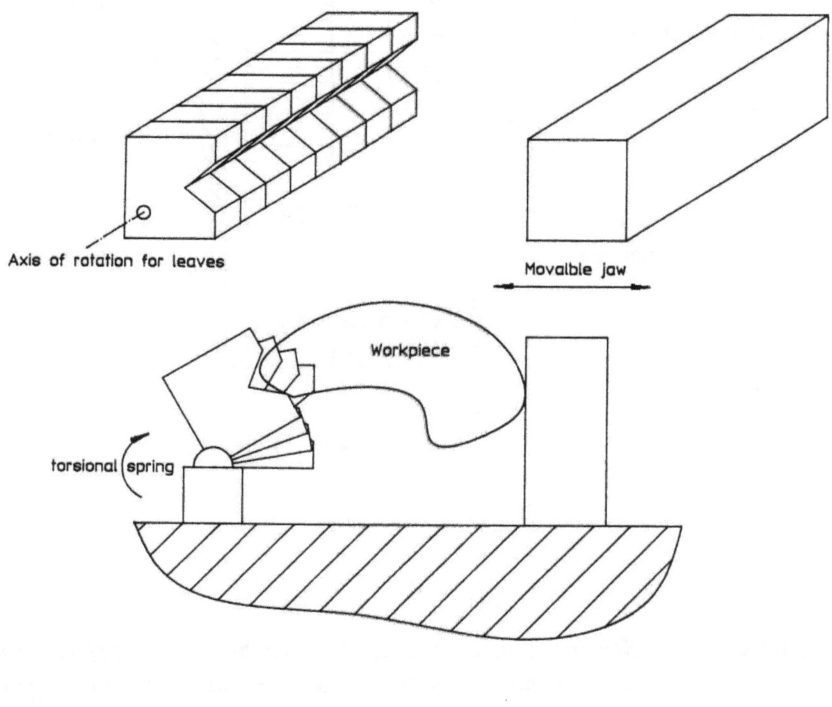

Figure 4.29 Multi-leaf vice [1] (Courtesy of ASME)

The petal collet is intended for clamping short asymmetrical components. It is made up of a number of 'petals' connected to a common base. The petals are closed by sliding the clamping ring upwards.

These devices are mechanically very simple but they are restricted in the components that can be successfully clamped. They are clamps only, location must be provided by separate mechanisms.

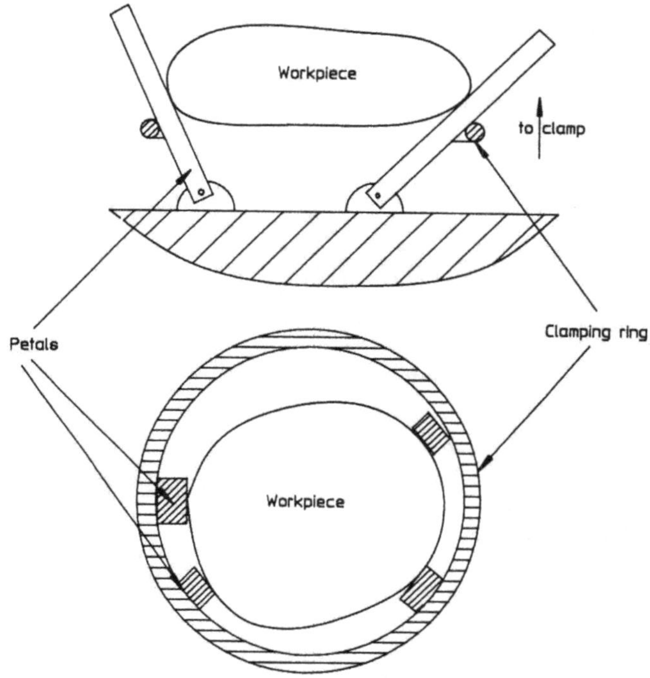

Figure 4.30 Petal collet [1] (Courtesy of ASME)

Englert and Wright have described an ingenious design for a conformable clamp which can accommodate a variety of parts such as turbine blades produced in a forging cell [20]. This clamping device can serve several hundred blade styles of different geometries. It consists of a hinged octagonal frame which can be manipulated to accommodate the blade and then be closed. The configuration of the programmable clamp is shown in Figure 4.31. This clamping device consists of a series of plungers located in the lower half of the clamp, activated by compressed air. When air pressure is applied, these plungers are pressed against the blade and subsequently conform to the profile of the blade. The blade is constrained by a high strength belt attached at the top half of the clamp and wrapped over the top surface of the blade holding the blade in position against the plungers. When the plungers have conformed to the blade shape, they are locked in place with socket screws to provide a mechanically stiff support for the blade. The air supply can then be disconnected. Two or three of these clamps can be used to hold the blade. The blade and clamp assembly can then be secured on a machine table.

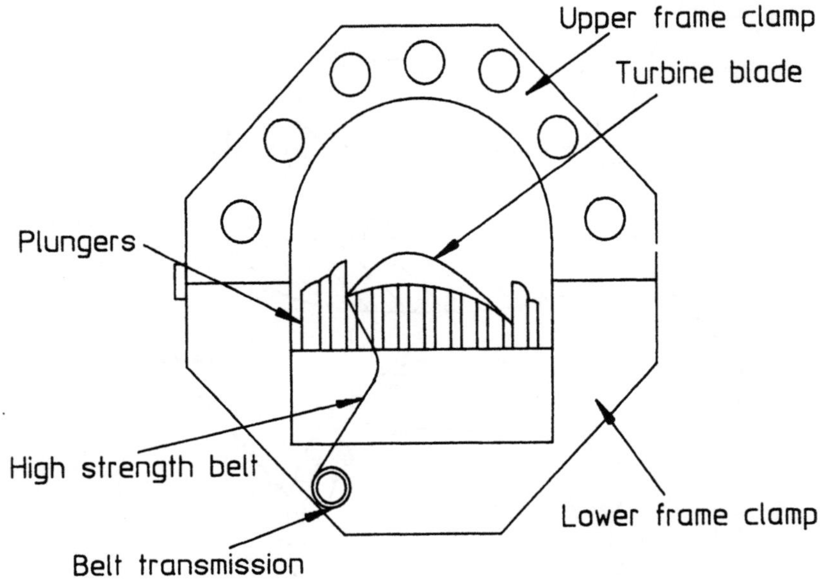

Figure 4.31 A turbine blade held in flexible clamps [1] (Courtesy of ASME)

- **Programmable conformable clamps**

  Wright, with Cutkosky and Kurokawa [21] have described an extension to the above method which enables the plungers to be actively modified under program control. The plungers are pre-configured to the correct profile using a master template. The master template consists of plungers that can be driven to the correct position using stepping motors. Since the master template is not part of the fixture it does not have to be compact. Unlike passive conformable clamps this technique provides location as well as support. Conformable clamps are however limited in the range of components for which they are suited.

4.3.5 Programmable Fixtures.

The ideal to which several researchers are striving is a method of constructing or reconfiguring fixtures completely automatically. Woodwark and Graham [10] were early leaders in applying robots to construct fixtures and more recently Benhabib,

Chan and Dai have described a programmable fixture system [22],[23]. The modular fixture elements discussed in the previous section lack "intelligence" in the form of sensory feedback capabilities and programmability. This is relevant when a fixture has to be built automatically using a robot. Some researchers have provided sensors at the baseplates and the elements so that feedback can be provided once an element is placed in position. The type of modular elements include locating pins, supporting pins, clamping elements and hole-based base plates. A hole-based system seems to be more efficient than a slot-based system as it is easy to locate the modular elements on the baseplate and keep track of the elements. Such modular elements can be specially designed for a specific purpose such as assembly [24], sheet metal drilling [25] and machining prismatic workpieces [26]. Benhabib [22] designed a modular programmable fixturing system consisting of :

- Horizontal locator
- Variable height vertical locator
- Variable width V-block
- Universal clamps
- Baseplates

All the components are sensor integrated, of variable height or variable width, and with standard component/baseplate interface. A horizontal locator developed by Chan and Benhabib [23] is shown in Figure 4.32. This system is capable of handling a wide variety of geometries thus ensuring re-configurability and flexibility.

4.4 Conclusions

In order to be cost effective and practical, the fixtures used in an FMS environment must be adaptable and reconfigurable to handle various part geometries. In this Chapter, an overview of the fixture technology and possible application in FMS was presented. Flexible fixtures such as modular fixtures, phase-change fixtures and programmable fixtures are capable of accomplishing the necessary objectives. On the other hand, dedicated fixtures tend to be more

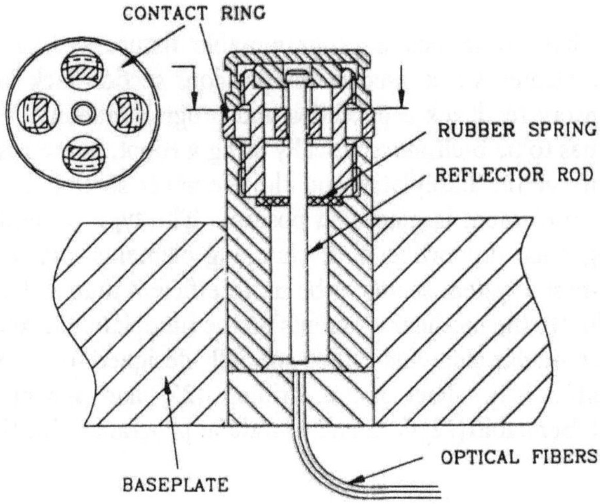

Horizontal Locator

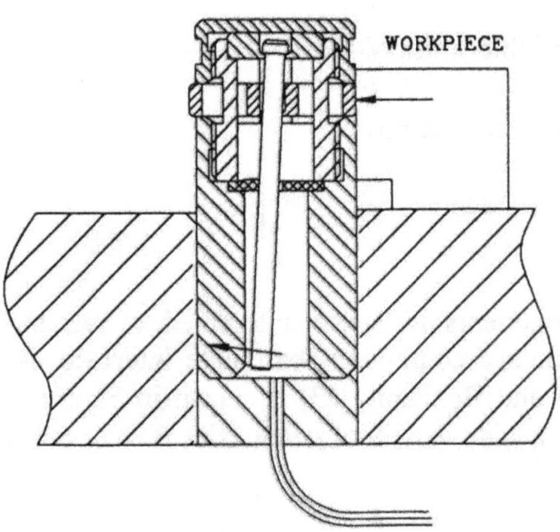

Horizontal Locator in Contact With a Workpiece

Figure 4.32 Horizontal locator (Reprinted with permission of the Society of Manufacturing Engineers [23] copyright 1990)

expensive as the product geometries change rapidly because there is a need to design a new fixture for each set-up. The next Chapter discusses the fixture design methodologies and the fixture design techniques using CAD.

4.5 References

1. Thompson BS. Flexible fixturing -A current frontier in the evolution of flexible manufacturing cells. ASME paper no. 84-WA/Prod-16. The American Society of Mechanical Engineers 1984.

2. Thompson BS, Gandi MV. A commentary on flexible fixturing. Applied Mechanics Review 1986; 39(10):1365-1369

3. AIOI Engineering Inc, 180 Bloor St.,West, Toronto, Ont., Canada

4. Boyes WE. Handbook of jig and fixture design. Society of Manufacturing Engineers, 1989.

5. Korsakov V. Fundamentals of fixture design. Mir Publishers, Moscow, 1989.

6. Blüco Technik. Modular Fixturing System-Parts Catalog 8/89, 1989

7 Hoffman GH. Modular fixturing. Manufacturing Technology Press, Lake Geneva, Wisconsin, 1987

8 Asada B, By AB. Kinematic analysis of workpiece fixturing for flexible assembly with automatically reconfigurable fixtures. IEEE Journal of Robotics and Automation 1985; RA-1(2):88-94

9. Ngoi BKA. Computer aided design of modular fixture assembly. PhD thesis, University of Canterbury, 1990

10. Neads SJ, Graham D, Woodwark JR. Towards a fixture building robot. UK Robotics Research 1984, I Mech E Conference Publications, London 1984, paper C470/84 : 99-104

11. Gandi MV, Thompson BS. Automated design of modular fixtures for flexible manufacturing systems. Journal of Manufacturing Systems 1986; 5(4): 243-252

12. Lewis G. Modular fixturing system. Second International conference on Flexible Manufacturing Systems, IFS (Conferences), London, 1983;451-462

13. Marcus A, Markusz Z, Farkus J, Filemon J. Fixture design using Prolog: an expert system. Robotic and Computer Integrated Manufacturing 1984; 1(2):167-172

14. Whybrew K, Ngoi BKA. Computer aided design of modular fixture assembly. Int J of Advanced Manufacturing Technology 1992; 7(5):267-276

15. Nee AYC, A Senthil kumar, S Prombanpong, KY Puah. A feature-based classification scheme for fixtures. Annals of the CIRP. 1992; 41(1) :189-193.

16. Gandhi, MV and BC Thompson. Phase-change fixturing for flexible manufacturing systems. J of Manu. Sys. 1985; 4(1):29-39.

17. Fisher Gauge Limited, P.O. Box 179, 194 Sophia St. Peterborough, Onterio, Canada

18. Perovskii AP, Perovskii SA. New elastic robot grippers. Soviet Engineering Research 1981; 1(7):10-11

19. Bidanda, B and CK Muralikrishnan. Flexible fixturing for intelligent Manufacturing.Intelligent Design and Manufacturing, Edited by Andrew Kusiak, John Wiley & Sons Inc, 1992.

20. Englert, PJ and PK Wright. Application of artificial intelligence and the design of fixtures for automated manufacturing. IEEE Int. conf. on robotics and automation, 1986:345-351.

21. Cutkosky MR, Kurokawa E, Wright PK. Programmable conformable clamps. AUTOFACT 4 Conference Proceedings. SME, Dearborn, Michigan,1985 11.51-11.58

22. Benhabib, B, KC Chan and MQ Dai. A modular programmable fixturing system. J of Engineering for Industry. 1991; 113:93-100.

23. Chan, KC, B Benhabib and MQ Dai. A reconfigurable fixturing system for robotic assembly. J of Manuf. Sys. 1990; 9(3):206-221.

24. Asada, H and A Fields. Design of flexible fixtures reconfigured by robot manipulators, Proceedings of ASME Winter Annual Meeting, Sym. on Robotics and Automation, 1985; 251-257.

25. Youcef-Toumi, K, WS Liu and H Asada. Computer-aided analysis of reconfigurable fixtures and sheet metal parts for robotic drilling. Proc. of the Int. Conf. on the Manufacturing Science and Technology of the Future, 1987.

26. Colbert, JL, R Menassa and WR CeVries. A modular fixture for prismatic parts in FMS. Proc.of North American Manufacturing Research Conference, 1986.

5 Computer-Aided Fixture Design

5.1 Computer-Aided Design of Fixtures

Fixture design has traditionally been regarded as a manual process due to extensive requirements of heuristic knowledge and skilled craftsmanship. There are many mathematical and scientific formulae which can be used to calculate cutting forces, deflection of structural members, tolerance analysis of locating datum, etc. However many of the good fixture design features such as ease of loading/unloading, safety considerations, ingenuity in securing the workpieces, etc come from the experience and skill of the designers. Due to the extensive requirements of heuristic knowledge and craftsmanship, the automation of fixture design has not been considered possible in the past. Recent developments in artificial intelligence techniques and, in particular, knowledge representation have provided great opportunities in automating this field.

5.2 Approaches in Fixture Design

As in computer-aided process planning (CAPP), the fixture design process can be approached using both variant and generative techniques. The proposed outline of the variant and generative fixture design schemes are presented in Figure 5.1 [1].

5.2.1 Variant Fixture Design

In variant fixture design, workpieces belonging to the same part family are assumed to have similar machining features and/or requiring similar operation sequences and set-ups. Fixture designers can often vary an existing design to quickly arrive at a similar design. In this approach, a fixture and workpiece classification and coding system has to be formulated. In addition to the information provided in the general

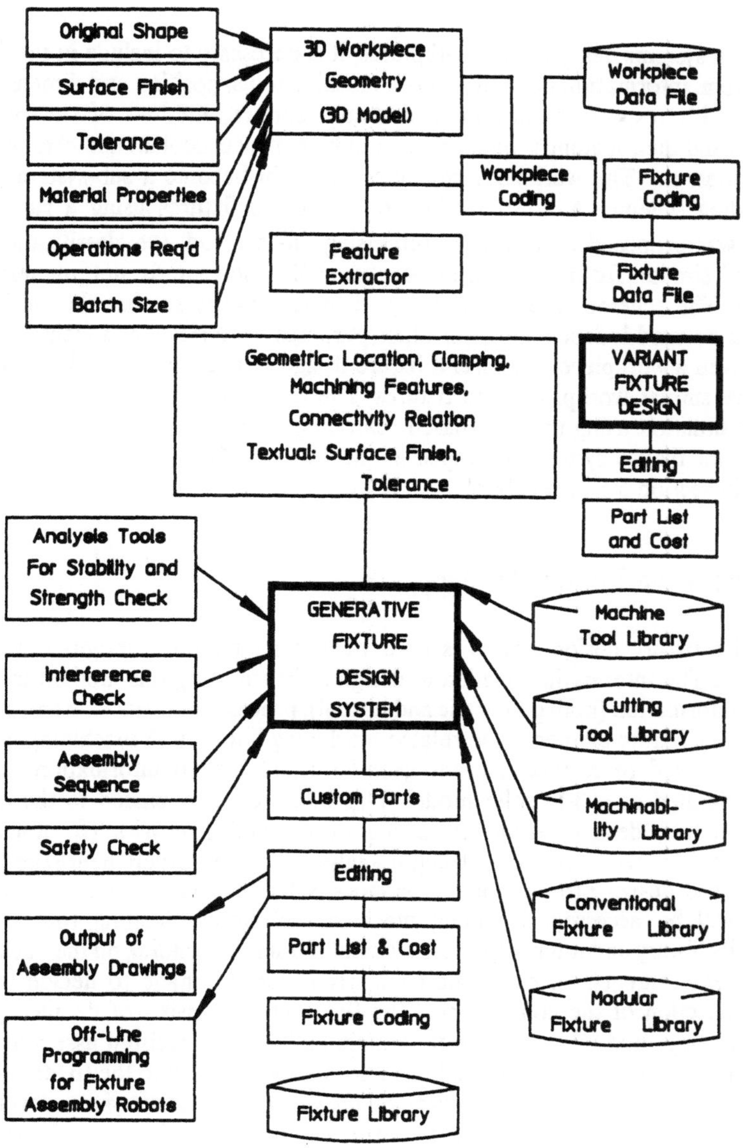

Figure 5.1 *Proposed outline of variant and generative fixture design systems [1]*

classification systems designed for workpieces, it is necessary to include in the fixture coding system, information on cutting tools, machine tools, position and dimensioning tolerances with respect to location and datum surface, method of securing the workpieces and their loading sequences, assembly sequence, coolant and chip disposal considerations, *etc.* The workpiece coding scheme will become a sub-system of the fixture coding scheme. A novel method of classifying fixtures using a rule-based technique was proposed by Nee *et al* [2]. Using this method, the classification and retrieval of similar fixtures is easier as the coding scheme is not numeric but descriptive. The schemes are to be designed in modularity so that the workpiece scheme can be used independently for other purposes of retrieval and identification.

Given a workpiece, the variant fixture design system must be able to identify and retrieve similar workpieces and fixtures. The retrieved fixture may be modified to suit the situation using the case-based reasoning approach. A detailed discussion on the variant fixture system developed at the National University of Singapore is explained in section 5.5 of this book.

5.2.2 Generative Fixture Design

A generative fixture design system is used when similar fixture designs cannot be retrieved [3]. The information needed to design a fixture using this method includes workpiece information (both geometric and textual), process plan, machine and cutting tool envelopes, fixture elements and related machining libraries. A machining feature recogniser is used to extract relevant geometric and textual information from a workpiece defined as a 3-D solid model. This information is passed to the expert generative fixture design system consisting of machining physics (formulae for evaluating cutting forces, stability, strength analysis, *etc*) and expert heuristics (rules of thumb, good design proportions, ergonomics, safety considerations, *etc*). Several databases will be accessed, including machine tool library, cutting tool library, machinability library and modular fixture element library. Depending on the batch size and other requirements, an economic comparison may be made to decide on the optimized selection of the components. The final output consists of detailed parts, assembly drawings and off-line robot programming for robot-assisted assembly. The newly created fixture will be coded and this information is added to the fixture library for future references. A detailed discussion on the generative fixture system developed at the National University of Singapore is explained in section 5.6 of this book.

5.3 Fixture Design Techniques

Fixture design techniques can be classified broadly into three categories based on the techniques used:

- CAD tools for fixture design
- Group technology concept for fixture design
- AI and expert system in fixture design
- Optimization of fixture configuration

5.3.1 CAD Tools for Fixture Design

Early applications of CAD techniques in designing fixtures merely used the draughting capabilities of a CAD package in producing drawings. Details of fixture elements such as pins, clamps and base plates are stored in the database and a user can pick and place them at any place as desired. This was found to be unsatisfactory as a large amount of library elements are needed and the effort required to generate them is formidable. In addition, only a particular supplier's components are stored in the library. This method reduces the amount of effort required in draughting the fixture elements. Although there is a reduction in the total time taken to produce a design, the final design depends largely on the designer's expertise.

5.3.2 Group Technology Concept for Fixture Design

Group technology is a manufacturing philosophy in which similar workpieces are grouped together to take advantage of their similarities in manufacturing and design [4]. The part code is a shorthand notation containing sufficient information about part characteristics so they can be conveniently retrieved. A workpiece and its fixture can be coded to contain information on workpiece characteristics, process planning and fixture layout. Similar fixtures for a given workpiece can be identified using the codes which can be retrieved and, if necessary, modified to meet the design requirements. This approach would reduce the design time and eliminate multiple designs of largely identical fixtures.

Miller and Hannam were probably the first researchers to suggest the use of a coding system to classify fixtures [5]. Instead of using numeric taxonomy, a new approach using an expert system was used to code fixtures [2] for storage and retrieval. This resulted in the development of a variant fixture design system. A detailed discussion on the variant fixture design system incorporating the GT concept is presented in section 5.5 of this book.

5.3.3 AI and Expert Systems in Fixture Design

Due to the capabilities of AI and expert systems, some researchers have started to explore the possibilities of using them for designing fixtures. AI can be defined as: "An intelligent computer program that uses the knowledge and inference procedures to solve problems that are difficult enough to require significant human expertise for their solution" [6]. Recently, this has become an essential tool to integrate CAD and CAM activities.

Fixture design rules and facts are stored in the knowledge base of an expert system. They are used to solve a domain-specific problem using an inference engine, also known as the reasoning mechanism. This approach resulted in the development of automated generative fixture design systems. A detailed discussion on the generative fixture design system using the AI techniques is presented in section 5.6 of this book.

5.3.4 Optimization of Fixture Configuration

Optimization techniques are an important tool in engineering design. The advantage of these techniques is the ability to determine the minima and maxima without exhausting all the possible solutions. Optimization techniques are mainly used in the fixture design context to identify the optimal fixture configurations such as the identification of locating, supporting and clamping points. Such techniques require an objective function incorporating the need and relevant constraints to satisfy the need so as to arrive at the optimal configuration.

These techniques are coupled with the finite element analysis to automatically arrive at the fixture configuration. In a finite element mesh only the nodal values give exact solutions. Thus in order to arrive at an accurate solution it is desirable to apply boundary conditions at the nodal points. The workpiece must be re-meshed to ensure that the support points obtained by the optimization routines are at the nodes of the finite element mesh. This process will optimize the fixture configuration to minimize the deformation of the workpiece under specific loading conditions. This reduces the need for manual redesign and results in reductions in the design lead time of the fixture.

5.4 Fixture Design Systems

Much research effort has been carried out to study the fixture design process using the above-mentioned techniques. Each approach has its own merits, shortfalls and usefulness. In general, the existing fixture design systems can be classified into three different categories based on their degree of automation, namely, interactive, semi-automated and fully automated. Some typical systems are reviewed in the following sections.

5.4.1 Interactive Fixture Design Systems

Interactive fixture design is a process where a computer is used to assist the designer by displaying the suitable fixture elements and the designer, based on his knowledge, decides on the fixture element to be used. The exact position of the fixture elements is decided by the designer in arriving at the final fixture configuration. Some of the systems that fall under this category are discussed below.

Various factors such as the choice of workpiece datum, geometrical form of the workpiece, the condition of the workpiece surface [7], material of the workpiece, machining operations [5], [8], [9] and [10] were considered while designing a fixture. However the designer selects the suitable fixturing faces, points and elements to build a fixture.

This method is time-consuming and the capabilities of computers are not fully exploited. Hence, further research was carried out and semi-automated systems were developed.

5.4.2 Semi-Automated Fixture Design Systems

A system can be said to be semi-automated if it does not require full knowledge or expertise from a designer while arriving at suitable supporting, locating and clamping faces, points and elements. Some of the systems that fall under this category are discussed below.

Ingrand and Latombe were probably the first researchers to develop a semi-automated system incorporating the expertise of a designer into the fixture design system [11]. The determination of suitable faces for supporting, locating and clamping can be decided automatically [12], [13]. However, the choice of suitable points and elements for building a fixture still depends on the user's expertise and knowledge.

Nnaji *et al* proposed a framework for a rule-based expert fixturing system for face milling planar surfaces using flexible fixtures [14]. Suitable points for supporting, locating and clamping in such systems are determined either using mathematical analysis of cutting forces [15], [16] or geometrical analysis [17].

5.4.3 Automated Fixture Design Systems

An automated system is one which obtains information directly from a CAD model and makes use of the knowledge base to decide on the suitable fixturing points and fixturing elements. Some of the systems that may partially come under this category are discussed below.

Design parameters such as orientation [18], stability [19], [20], [21], [22] [23], deflection due to cutting forces [24], set-ups [25], [26], tolerance relationships [27], assembly and interference [28] were considered while designing a fixture.

Expert systems [18], [29] together with a good knowledge representation scheme [30; 31] were generally used in arriving at a fixture design. Several other attempts were made by researchers to automate the assembly of fixtures using robots [32], [33], [34].

Figure 3.2 summarises some of the published fixture design systems and their major characteristics. It is apparent that the various reported automated fixture design systems are incomplete in one way or another [39].

5.5 A Variant Fixture Design System Using GT

As mentioned earlier, workpieces belonging to the same part families are assumed to have similar machining features and/or requiring similar operation sequences and set-ups. To be able to carry out a variant fixture design, it is necessary to have a suitable coding and classification scheme where the workpiece, processing environment, machine tools, cutting tools and the fixture can be sufficiently and accurately described. Numerical taxonomy will be difficult, if not impossible, to provide a full description of the information required. A coding system based on structural description is developed and discussed in this section. Contrary to most of the established coding and classification systems which use numerical taxonomy, the proposed workpiece and other schemes are classified according to goal-directed structured descriptions. In using the numerical taxonomy, it is difficult for the user to compare the best alternative classification and the description-numerical-description conversion is error-prone and difficult to interpret.

ATTRIBUTES	Systems	AIFIX (1986)	CADF (1986)	PBFIS (1989)	CAD MFS (1987)	JIGS (1985)	NEE (1987)	BOPAYA HANAPA (1989)	CHOU (1986)	GANDHI (1986)	MOD BUILD (1984)	HOLDEX (1986)	AUTOFIX (1990)	OFDS (1990)	Kumar (1992)	Ngoi (1990)	S-Anuwal (1992)
SYSTEM METHOD	INTERACTIVE				×	×	×										
	SEMI_AUTO			×				×			×						
	AUTOMATED	×	×						×	×		×	×	×	×	×	×
DESIGN APPROACH	GT METHOD				×	×											
	RULE_BASED		×				×	×	×		×				×	×	×
	EXPERT SYS	×		×						×		×	×		×		
PART SHAPE	PRISMATIC	×	×	×	×	×	×		×		×	×	×	×	×	×	×
	ROTATIONAL				×	×		×									
PART GEOMETRY	2_DIMENSIONAL		×		×	×	×	×									
	3_DIMENSIONAL	×		×					×	×	×	×	×	×	×	×	×
DESIGN CONSIDERATION	STABILITY	×	×	×					×	×				×	×	×	×
	COST	×	×														
	DEFLECTION		×	×										×	×		×
	MACHINED AREA			×			×							×	×	×	×
	POSITIONAL ACCURACY						×										×
	ASSEMBLY															×	
CAD INTEGRATION			×		×	×	×	×	×	×	×	×	×	×	×	×	×

Figure 5.2 Current state of computerised fixture design systems

An overview of the developed variant fixture design system, as shown in Figure 5.3, consists of coding, classification and design schemes. In general, the developed system consists of various modules such as:

- Case representation
- Case indexing
- Case retrieval
- Case modification

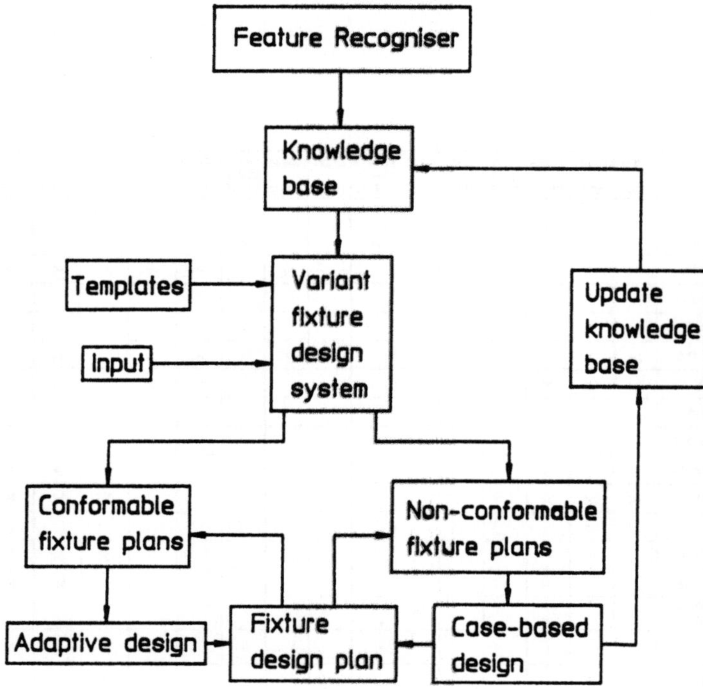

Figure 5.3 An overview of the developed variant fixture design system

A detailed discussion on the system developed at the National University of Singapore is presented in the subsequent sections of this Chapter.

5.5.1 A Feature-Based Classification and Coding Scheme

The feature-based classification system consists of three schemes,

- the workpiece coding scheme
- the fixture element coding scheme
- the fixture classification scheme

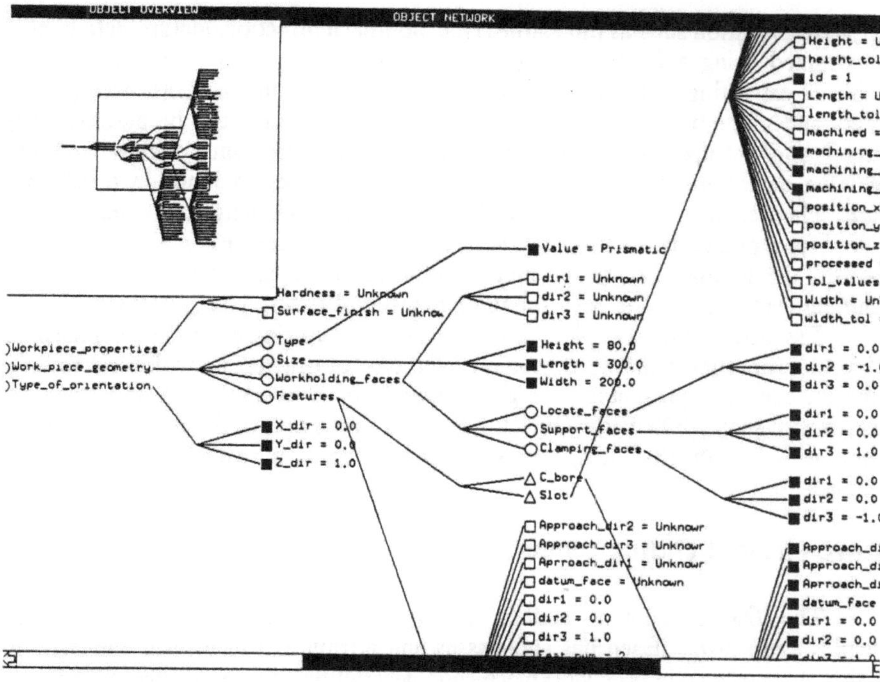

Figure 5.4 Partial list of workpiece coding scheme

Each fixture is classified based on the above coding schemes and represented as a case in the case-base. The schemes use goal dependency network and incorporate background knowledge which include classification goals, classification evaluation criteria as well as deductive and inductive inference rules [35], [36].

Workpiece coding scheme

Information such as the feature type, approach direction, feature details, etc., are extracted using a feature recogniser [3]. The developed feature recogniser is capable of extracting some 30 different types of features such as slots, holes, steps, pockets and a combination of theses features. These features can be automatically grouped into set-ups based on machining directions, tolerance factors and the machining environment and each set-up is assumed to be machined in a different fixture. Theoretically, in a 3-axis machining centre environment, six set-ups are required for a prismatic workpiece if there are machining features on all the six faces. However, the number of set-ups may be reduced by using either a 4- or 5- axis machining centre hence the corresponding number of fixture configurations.

Essentially, the workpiece code (goal), with sub-classes (sub-goals) such as workpiece properties, workpiece geometry and type of orientation with their various attributes, can be represented in a goal dependency network formulated using an expert system shell NEXPERT_Object. Figure 5.4 shows part of the workpiece coding scheme with known attributes, shown in dark squares, derived from the feature recogniser.

Fixture element coding scheme

The fixture element coding scheme consists of elements, operational principles and set-ups. Each has sub-classes and attributes which form connections with the workpiece code. The workpiece code descriptions, when matched with those in the fixture code, generate attributes that belong to the various sub-classes of the fixture element code, producing descriptions of the fixture elements. The partial list of the fixture element coding scheme is shown in Figure 5.5.

Fixture classification scheme

Fixtures commonly designed for prismatic workpieces can be classified according to location, clamping and operational principles. Each sub-class will have its own attributes which form the connections with the fixture element coding scheme, and in turn, with the workpiece coding scheme. Typical fixture configurations for prismatic parts are shown in Figure 5.6.

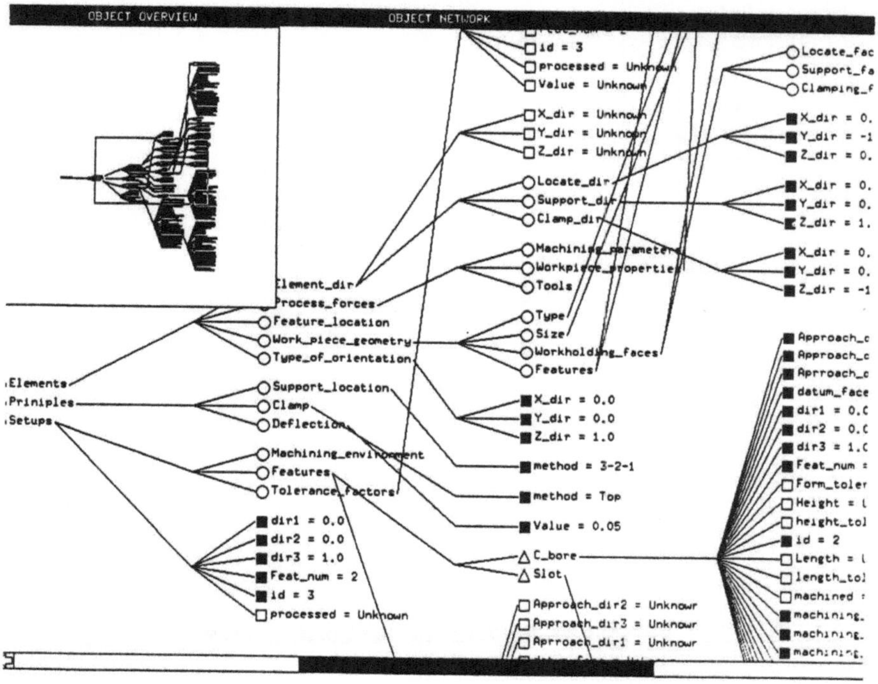

Figure 5.5 Partial list of the fixture element coding scheme

5.5.2 Case Indexing

The ability to retrieve and modify similar cases from the case-base to suit a new situation greatly depends on how each case is indcxed. The indexing should be simple,

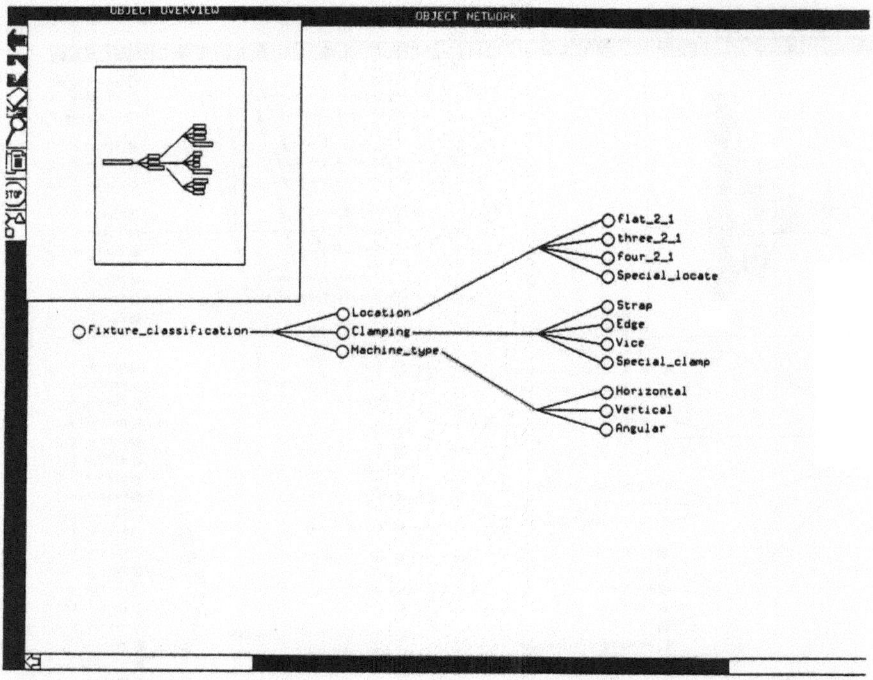

Figure 5.6 The fixture classification scheme

accurate and efficient. A fixture case can be indexed by considering the following parameters.

- *Workpiece characteristics*
  Workpiece characteristics consist of the features present, orientation of the workpiece and its geometry.

- *Locating principles*
  Locating principles used include 3-2-1, 4-2-1, locating from holes, cylinders, etc.

- *Clamping methods*
  Clamping methods considered include edge and top clamps, vices, etc.

- *Processing details*
  Type of machining operations involved are milling, drilling, boring, etc.

- *Machining environment*
  Machining environment includes 3-, 4-, 5- axes horizontal and vertical machining centres.

5.5.3 Case Retrieval

A 3-D CAD model is given as an input to the system. The features present in the CAD model as well as the non-geometric information associated with the features are extracted automatically via a feature recogniser. The number of set-ups required is determined and coded, based on rules, into the expert system network which forms a template for each set-up. The descriptions are then defined in the fixture element coding scheme for association with the various fixture elements which will be used and in turn, matched with the classes of fixtures which have been pre-defined for prismatic workpieces. This process is shown schematically in Figure 5.7.

This process can retrieve similar workpieces along with their associated fixtures stored in the form of templates. The case which matches the maximum number of attributes of the indexing parameters is retrieved. In order to reduce the search space, databases of workpieces are grouped and classified as prismatic, rotational, special, etc. The workpiece type is considered a significant "seed" to classify and group the database so as to reduce the search space. A simple rule depicting the process of grouping is shown below:

```
IF      ( is_equal ( workpiece_geometry.value,"prismatic" ) )

THEN  ( create_object ( temp_object, temp_type )
        ( let_grouping_be_true ) )
```

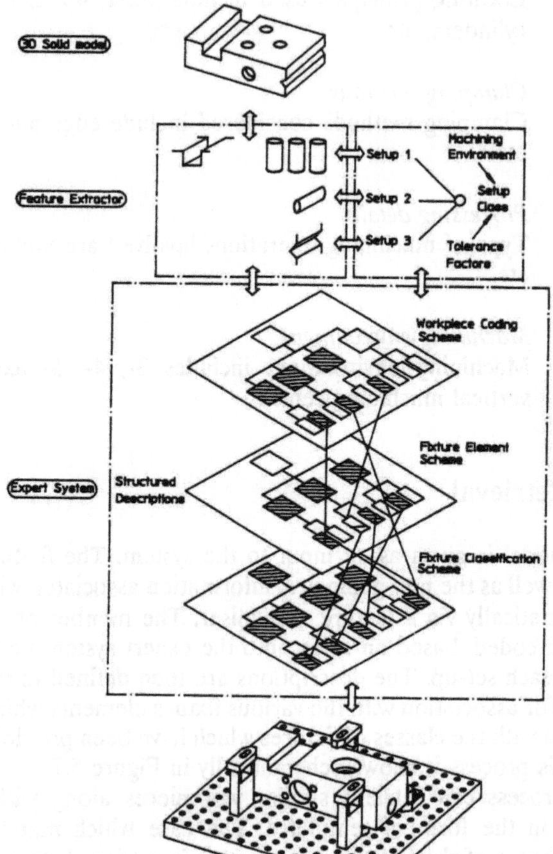

Figure 5.7 ***Schematic connectivity of the three schemes in arriving at a similar fixture design***

Further search is carried out in the new group so greater efficiency can be achieved. The other significant seeds used to classify the workpiece category are features, directions and sizes. Finally, an existing workpiece similar to the one in question is retrieved together with the corresponding fixture descriptions which match the descriptions of the fixture element code. The retrieved fixture is then classified to arrive at the nearest possible fixture, lower diagram of Figure 5.7.

5.5.4 Case Modification

The purpose of case modification is to modify the retrieved case to suit the needs of the present situation. Given below are the two possible types of situations that can result after a case is retrieved.

- The retrieved case is exactly the same as the present case and only requires relocation of fixture elements to suit the new situation. Such a case is known as a conformable fixture plan.

- The retrieved case has additional or fewer features that may require new fixture components or fixture configuration. Such a case is known as a non-conformable fixture plan.

Conformable fixture plan

The fixture configuration of the retrieved case can be modified either manually or using rules defined in the expert system. The similarity of the workpiece to the retrieved case can be identified by comparing the shape, size and the relative orientation of the features on the body. The rules defined can be used as guide lines to identify the supporting, locating and clamping points which may differ from the retrieved case based on the information obtained for the retrieved case. This information can be used to position and assemble the modular elements appropriately for the current workpiece.

Non-conformable fixture plan

In addition to identifying the supporting, locating and clamping points, work has to be done in selecting the fixture elements or reconfiguring the fixture based on the new workpiece. This is done using the rules embedded in the expert system.

5.6 A Generative Fixture Design System Using an Expert System

A generative fixture design system requires an intelligent feature recogniser, a sophisticated design system, a good knowledge representation scheme and sound interfacing of these modules. The integration of all the modules can only be achieved if there are sub-systems that can utilize design data from a CAD modeller and knowledge bases containing rich manufacturing and fixture information. Such an integration is not possible without a proper knowledge-representation scheme.

The architecture of the developed Modular Element Fixture Design Expert System (MEFDES) is shown in Figure 5.8 [37],[40]. The first module deals with designing a workpiece and attaching tolerance and surface finish as attributes. In the second module, the machining features are recognised and suitable modular elements are selected from the database to design a fixture. The third module, the vision system, deals with automatic assembly and inspection of the designed fixture. The vision system is still under development and not covered in this book. In general, the MEFDES can be divided into two sub-systems: the CAD sub-system and the expert sub-system. The two sub-systems interact with the three functional modules in arriving at a fixture design. In the CAD sub-system, a workpiece is designed using standard CAD routines and the machining features are extracted automatically by means of the feature recogniser. The expert sub-system is used to design a fixture using rules defined in the system. The workpiece design module makes use of the CAD sub-system to design the workpiece, whereas the fixture design module makes use of both CAD and expert sub-systems to design the fixture. This is explained in detail in the subsequent sections of this chapter.

5.6.1. Workpiece Design Module

The structure of the workpiece design module is shown in Figure 5.9. In MEFDES, the user is required to provide only geometric entities, *i.e.*, without taking into account the manufacturing details. Tolerance and surface finish will be attached as attributes to the CAD model at a later stage. The workpiece in the current system is represented by means of its faces, as they form the basis for fixture configuration. A typical face frame is as follows:

```
(Face                      /* Face of a part */
        (Dir1              /* X direction of normal vector */
         Dir2              /* Y direction of normal vector */
```

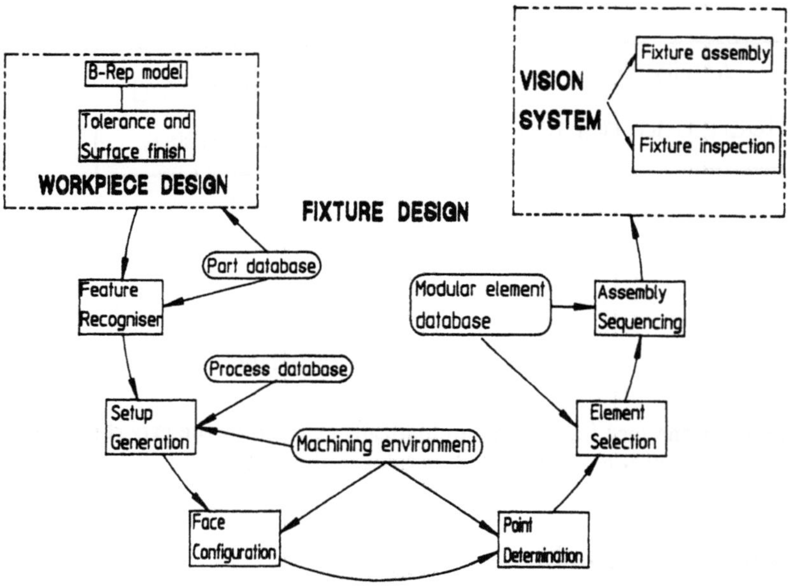

Figure 5.8 Detailed system architecture

```
Dir3                /* Z direction of normal vector */
Existed             /* If the face has been machined */
Method              /* Fixturing method */
Name                /* Face name */
Number              /* number of fixturing points */
Selected            /* If selected for fixturing */
Surface_finish      /* Surface finish */
Tolerance))         /* Specified tolerance */
```

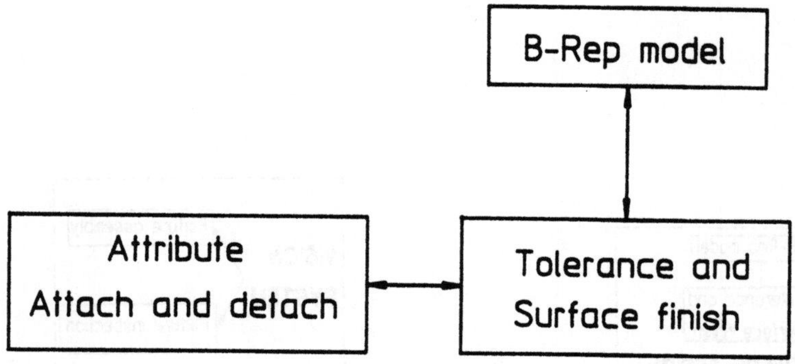

Figure 5.9 Workpiece design system

The slots shown in the face frame can automatically be filled by the expert sub-system using procedural knowledge represented as rules in the system.

5.6.2 Fixture Design Module

The fixture design module consists of six sub-modules:

- Feature extraction,
- Set-up generation,
- Face configuration,
- Point determination,
- Element selection and assembly sequencing

Each sub-module will be discussed in the subsequent sections of this chapter.

Feature extraction

The first step in fixture design is to identify the machining features in the workpiece. The developed system uses a graph-based approach which explicitly represents the connectivity relationship between any two faces in the Attributed Adjacency Matrix (AAM).

Three steps are involved in identifying a feature. They are:

- feature classification
- feature identification
- approach direction determination

Feature classification is the process of classifying features in their generic forms based on their topology, *viz.*, slot, step, *etc*. A specific feature is then identified in a selected generic group using rules and geometries of the feature. For example, a generic slot may be a dovetail, an inverted dovetail or a right-angled slot. A specific feature is further classified based not only on its topology but also its geometry, *i.e.*, dimensions as well as the angular definition of the faces, *etc*. The approach direction, *i.e.*, the unobstructed direction of the approach of a cutting tool, is determined by considering the orientation of a feature. The approach direction of a blind hole is the vector normal to the machining face, which is the bottom face (planar) of the blind hole. A complete list of the recognisable machining features is listed below.

- Slot
- Blind Slot
- V-Slot
- Blind V-Slot
- T-Slot
- Blind T-Slot
- Cylindrical Slot
- Blind Cylindrical Slot
- Through Hole / Hole
- Blind Hole
- Tapered Hole
- Blind Tapered Hole
- Triangular Hole
- Blind Triangular hole
- Multi-Facet Hole
- Countersunk
- Blind Countersunk
- Counterbore
- Step
- Blind Step
- Slant Step
- Dovetail

- Blind Dovetail
- Inverted Dovetail
- Blind Inverted Dovetail
- Chamfer
- Obtuse Chamfer
- Pocket
- Island
- Boss

The information extracted from the CAD model is represented in the NEXPERT_Object as frames. A typical feature frame is as follows:

```
Feature              /* Feature of a part */
      (Dir1          /* X-axis approach direction */
       Dir2          /* Y-axis approach direction */
       Dir3          /* Z-axis approach direction */
       Machined      /* If machined or not */
       Position_X    /* X reference location */
       Position_Y    /* Y-reference location */
       Position_Z    /* Z-reference location */
       Length        /* Length of the feature */
       Width         /* Width of the feature */
       Depth         /* Depth of the feature */
       Length_tol    /* Feature length tolerance */
       Width_tol     /* Feature width tolerance */
       Depth_tol     /* Feature depth tolerance */
       Name          /* Feature name */
```

Set-up generation

The first step in the set-up generation sub-module is to group all the features to obtain three clusters: tolerance factor cluster, face abnormalities cluster and machining direction cluster.

- *Machining direction cluster*
  The features having the same machining direction are grouped together to form clusters. Typically, a prismatic workpiece machined on a 3-axis vertical machining centre can have up to 6 clusters *i.e.*, one cluster for each face.

- *Face abnormalities cluster*
  From the fixturing point of view, it is advisable to use the maximum projected area for supporting, locating and clamping. This will ensure maximum stability and rigidity while machining a workpiece. If the workpiece supporting area is too small, it may be difficult to hold the workpiece rigidly. Hence, a face which projected area (ratio of the total area to the features area) is smaller, due to the presence of features, is generally to be machined at a later stage than the one which area is greater. The clusters are sequenced in a descending order based on the projected areas.

- *Tolerance factor cluster*
  Features machined must be held within specified tolerance relationship with reference to datum surfaces, within or between features. Tolerance violation may be due to misalignments from rotation and/or translation of a workpiece. In most cases, errors due to rotation are dominant and are difficult to control, whereas errors due to translation can be controlled relatively easier. Hence, consideration of rotational errors is important in the set-up generation phase. In order to compare different types of tolerances, the so-called tolerance factor, which is a non-type-specific value, has been introduced. It is a measure of the relative importance of the tolerance variations between features [25],[38]. In the case of parallelism, the tolerance factor is evaluated by dividing the permissible error by the longest diagonal of the toleranced face. Machining precedence is formulated by considering the order of the tolerance factor of the features.

Using these clusters and the rules established in the expert sub-system, the set-up sequence can be decided automatically. A typical rule to arrive at a set-up sequence is shown below:

```
IF (        any feature present in one node appears in any other
            node in a cluster,
            AND, that feature is present alone in a cluster
  )
THEN (      delete that node from that cluster
            LET cluster refinement be TRUE
    ))
```

```
IF (            any of the features are present in all the nodes of a
                cluster,
                AND, the total number of nodes in that cluster > 2,
   )
THEN (          feature to be removed from a node depends on the
                other two clusters
     )
```

A typical frame for a set-up is as follows:

```
(Set-up                      /* Set-up for machining */
        (Dir1                /* Machining direction X */
        Dir2                 /* Machining direction Y */
        Dir 3                /* Machining direction Z */
        Feat_num             /* Number of features in that set-up */
        Id                   /* Id of that set-up */
        Processed ))         /* If machined or not */
```

Face configuration

Depending on the set-up selected, the workpiece information is updated automatically in the expert sub-system, *i.e.,* if set-up 2 is selected for designing a fixture then set-up 1 must be completed.

The face configuration sub-module determines the suitable faces for supporting, locating and clamping. The face configuration has three sub-classes, namely, supporting face class (BFList), locating face class (SFList), and clamping face class (CFList). The locating face class is further divided into major locating class (SList1) and minor locating class (SList2). The suitable faces are attached to these classes dynamically and their properties such as area, machining directions, *etc.*, are automatically inherited from the face class. The suitable faces for locating, supporting and clamping are automatically decided using the rules established in the expert sub-system.

Determination of supporting, locating and clamping points

The exact points for supporting, locating and clamping the workpieces are determined in this sub-module. These are the points where the modular elements will be in contact with the workpiece. Rules to determine these points have been established in the expert sub-system. Supporting points are attached to the supporting point class "BFPoints", locating points are attached to the locating point class "LFPoints" and clamping points are attached to the clamping point class "CFPoints". The co-ordinates of the points are computed and attached to the properties of the point class using NEXPERT commands. A general point frame is as follows:

```
(Points
        (Name              /* Point name */
        X                  /* X coordinate */
        Y                  /* Y coordinate */
        Z                  /* Z coordinate */
        Existed ))         /* If modular tower is built or
                           not */
```

Selection and assembly of modular fixture elements

The suitable modular elements for supporting, locating and clamping are automatically selected from the modular element database. Once an element is selected, the inventory status of the elements present in the database is updated. The clamping elements are selected based on the clamping height which is the sum of supporting element height and the workpiece height. The final state of the object network for the entire system is shown in Figure 5.10. From Figure 5.10, it can be seen that "Faces class" consists of face objects which are linked to the "BFList class" or "CFList class" or "SFList class". The "Feature class" has features which comprise of faces and hence the "Face class" is linked to the "Feature class" through the faces. "Set-up class" consists of set-up objects with features and the "Set-up class" is linked to the "Feature class". Similarly, the "Point class" consists of points which fall on the faces and it is also linked to the "Modular element class". This shows how the various classes and objects are inter-linked and information inheritance from one another can be achieved while designing a fixture. A typical representation of a modular element frame is as follows:

```
(Mod_element           /* Modular element */
    (Function          /* Function of modular element */
     Height            /* Height of element */
     Length/diameter   /* Circle length or diameter */
```

```
Name            /* Name of the element */
Num_select      /* Inventory on selection */
Range           /* Range of the element */
Selected        /* Current status of element */
Type            /* Element type-rigid or adjustable */
```

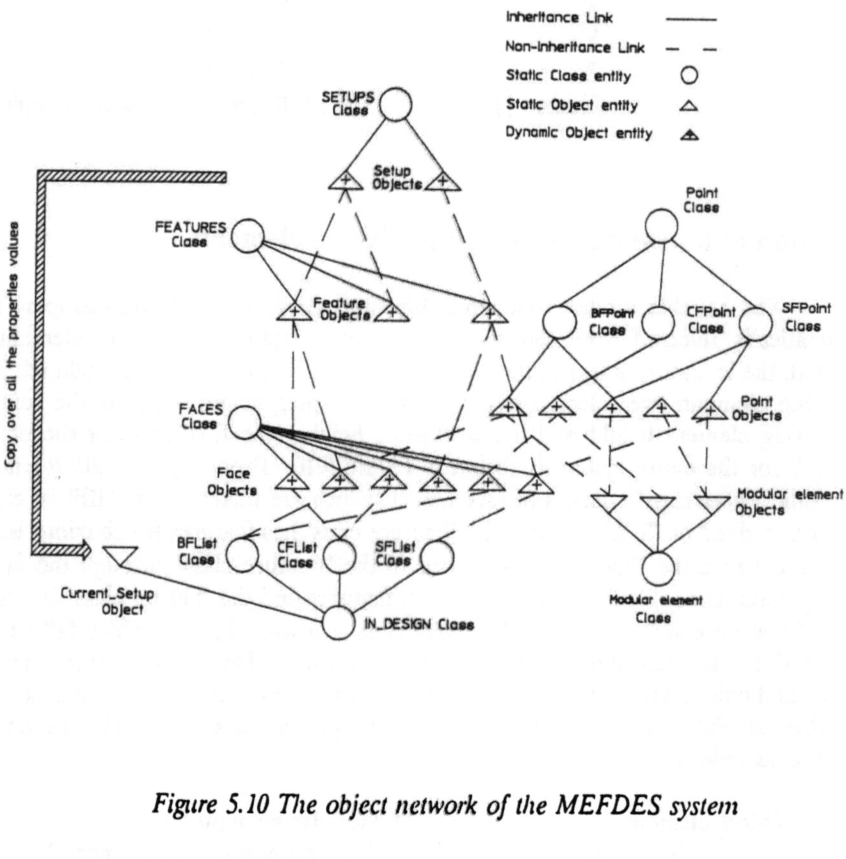

Figure 5.10 The object network of the MEFDES system

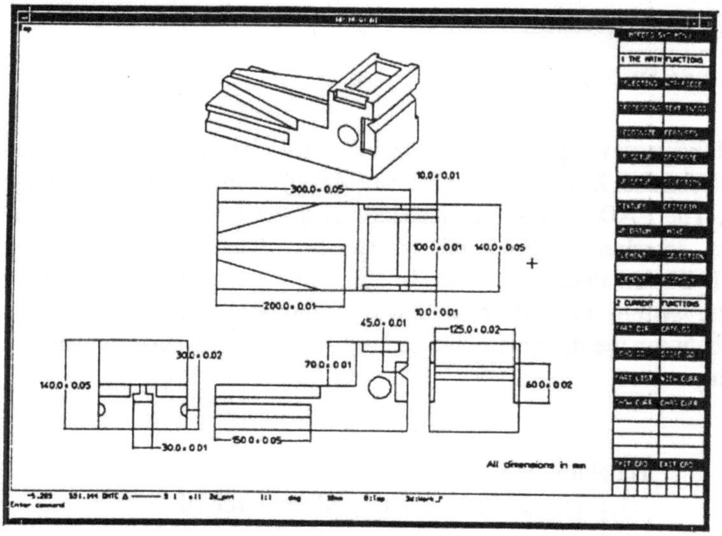

Figure 5.11 An example workpiece

5.6.3 Case Studies

In this section, an example showing the capability of MEFDES is presented. It is capable of designing fixtures for both horizontal (HMC) and vertical machining centres (VMC). As discussed above, MEFDES can be used to design fixtures for a selected set-up or for all the set-ups of a workpiece.

An example workpiece shown in Figure 5.11 is to be machined on a VMC. Figure 5.12 gives the output of the sub-system showing the recognised features, their machining faces and faces forming the features. The output of the set-up generation sub-module is shown in Figure 5.13. Based on the above information, suitable modular elements are selected from the modular element database for the final assembly. Figures 5.14 and 5.15 show the assembled fixture with and without the workpiece.

5.6.4 System Implementation

The MEFDES is developed on a Hewlett Packard (HP) Series 360 workstation. It uses the Mechanical Engineering series 30 (ME30) solid modeller for designing the workpiece. The CAD database is decoded using the routines developed using the Application Interface (ME30 AI) software. The developed system uses NEXPERT_Object, an expert system shell for defining the fixture design knowledge.

The system integration is achieved through a complex mechanism of interfaces between the design module and the programming environment. It includes the following:

- Database interfacing
- Interface with the various modules involved

The interfacing mechanism is developed using C routines and linked to the NEXPERT's frames. When the program is executed, data such as faces, and their area, directions, etc., are computed using C routines by interacting with a CAD data structure. The set of data obtained form the CAD model is then sent to NEXPERT for further processing. The modular element database and NEXPERT are integrated using the C routines developed. Interfacing with different modules is achieved by two means: frames in NEXPERT and C data structures. The features extracted from the CAD model are transferred to the NEXPERT's knowledge base using the pointers. These pointers are used simultaneously to interact with the CAD model whenever required, thus allowing the complete system to be used in an automated manufacturing environment. Since the ME30 CAD modeller only allows a single line display for inputs, X Window is used to enhance user interface with the system.

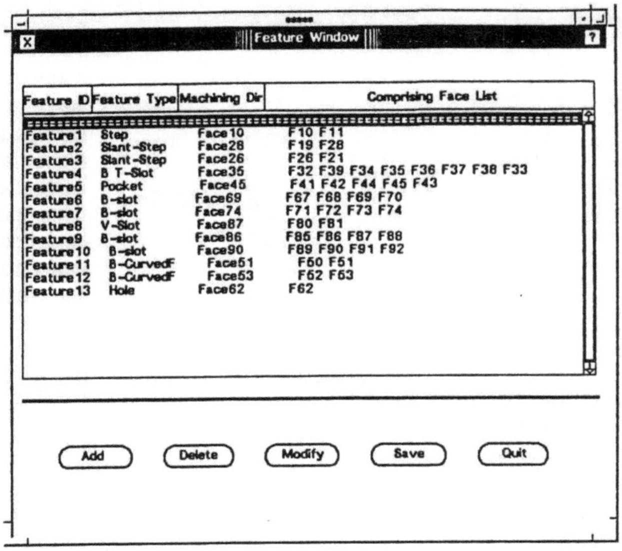

Figure 5.12 Feature window showing the recognised features, machining directions and the comprising face list

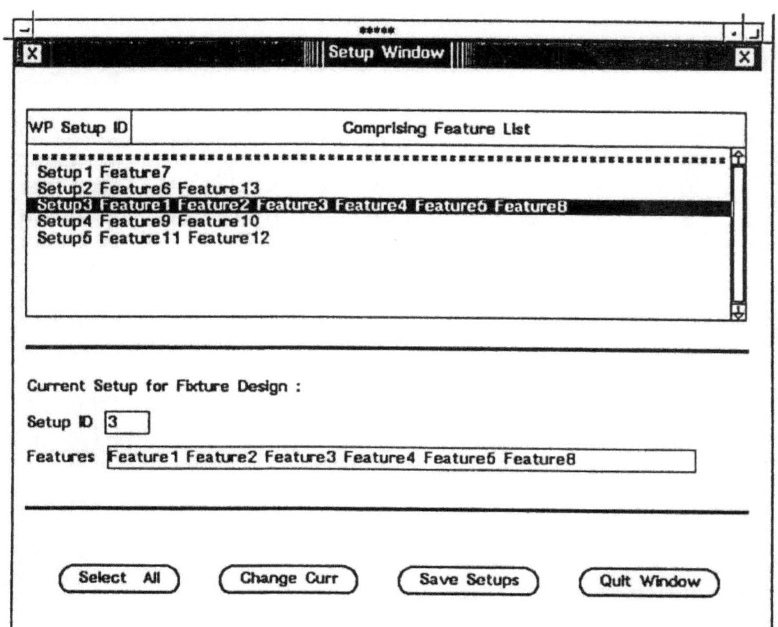

Figure 5.13 Set-up window showing the generated set-ups

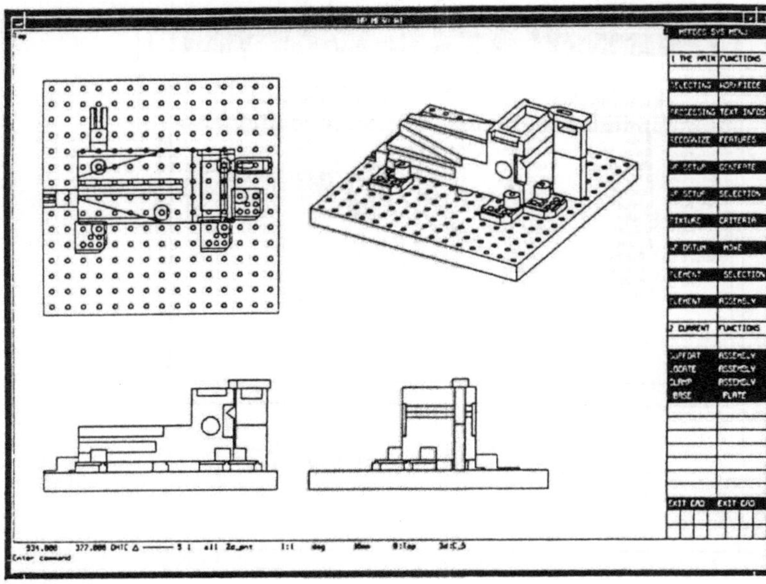

Figure 5.14 Assembly of selected elements with the workpiece

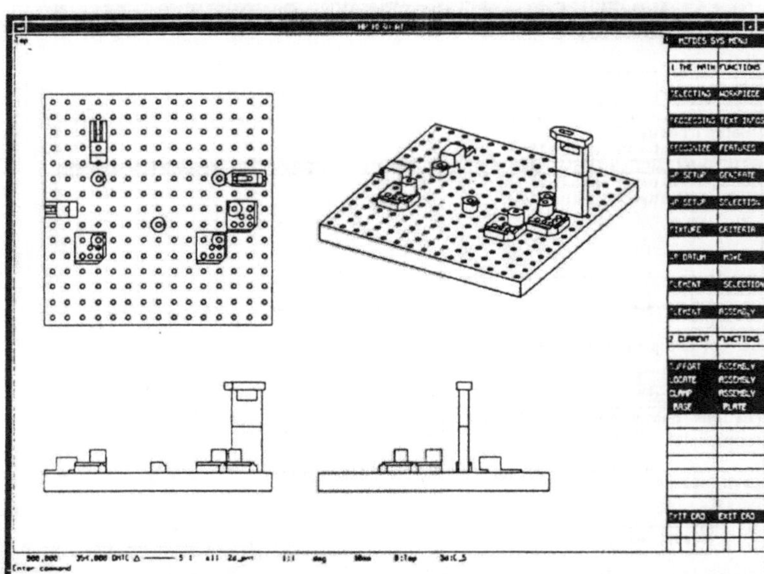

Figure 5.15 Assembly of modular elements without the workpiece

5.7 References

1. Nee, AYC and AN Poo. Expert CAD systems for jigs and fixtures, Artificial Intelligence in Design (edited by D.T. Pham), IFS/Springer. 1990: 343-370

2. Nee, AYC., A Senthil kumar, S Prombanpong, and KY Puah. A feature-based classification scheme for fixtures, Annals of the CIRP. 1992; 41(1):189-193

3. Nee, AYC and A Senthil kumar. A framework for and Object/Rule-based automated fixture design system, Annals of the CIRP. 1991; 40(1):147-151

4. Groover, MP and EWJR Zimmers. CAD/CAM: Computer-Aided Design and Manufacturing. Prentice-Hall Inc., New Jersey, USA. 1984.

5. Miller, AS and RG Hannam, Computer aided design using a knowledge base approach and its application to the design of jigs and fixtures. Proc. I.Mech.E. 1985; 199(B4)

6. Cohen, PR and E Feigenbaum. The Handbook of Artificial Intelligence, William Kaufmann Inc., Los Aitos, California, USA. 1982.

7. Imhof, G and W. Grahl. Die auswahl von vorrichtungselementen als bestandteil der rechnergestutzten projektierenden konstruktion von vorrichtungen, Wiss. Z. d. Techn. Hochsc. Karl-Marx-Stadt. 1977; 19(1):49-61

8. Nee, AYC, N Bhattacharya, and AN Poo. Applying AI in jigs and fixtures design. Robotics and Computer Integrated Manufacturing. 1987; 3(2):195-201

9. Grippo, PM, MV Gandhi and BS Thompson. The computer aided design of modular fixturing systems. The Int. J. of Adv. Manuf. Tech. 1987; 2:75-88

10. Puah, KY and AYC Nee. A microcomputer-based modular fixturing system. Proc., Asia-Pacific Industrial Automation (IA) '90 Conference on Automation: Strategies for Manufacturing Growth, Singapore. May, 1990:580-592

11. Ingrand, F and J Latombe. Functional reasoning for automatic fixture design. Lifia (IMAG) France, 1980.

12. Markus, A, Z Markusz, J Farkas and J Filemon. Fixture design using PROLOG: An expert system. Robotics and Computer Integrated Manufacturing. 1984; 1(2):167-172.

13. Bopaya Monapa, B. An integrated CAD-CAM approach for the selection of workholding devices for concentric, rotational components. PhD Dissertation, The Pennsylvania State University, USA. 1987.

14. Nnaji, BO and P Lyu. Rules for an expert fixturing system on a CAD screen using flexible fixture. J of Intelligent Manufacturing. 1990; 1:31-48.

15. Cadabaj, J. Theory of computer aided fixture design. Computers in Industry. 1990; 15:141-147

16. Mittal, RO, PH Cohen, and BJ Gilmore. Dynamic modelling of fixture-workpiece system. Robotics and Comp. Integrated Manufacturing. 1991; 8(4):201-217

17. Shin, H. Rule-based systems approach to fixture design. PhD Dissertation, Dept. of Industrial and Systems Engineering, The Ohio State University, Ohio, USA. 1988.

18. Ferreira, PM, B Kochar, CR Liu and V Chandru. AIFIX: An expert system approach to fixture design. ASME Computer Aided/Intelligent Process Planning. 1986; PED-Vol.19:73-82

19. Chou, CY. Computerized fixture design from solid models of workpieces. Master Thesis, School of Industrial Engineering, Purdue University, West Lafayette, Indiana, USA. 1985

20. Gandhi, MV and BS Thompson. Automated design of modular fixtures for flexible manufacturing systems. Journal of Manuf. Sys. 1986; 5(4): 243-252

21. Prombanpong, S. A computer aided fixture design system for two dimensional prismatic parts produced on machining centres. Masters Thesis, Dept. of

Industrial and Systems Engineering, The Ohio State University, Columbus, Ohio, USA. 1986

22. Lim, BS and JAG Knight. Holdex - Holding device expert system. Proc. 1st Int. Conf. on Applications of Artificial Intelligence in Engineering Problems 1986:483-493

23. Englert, PJ and PK Wright. Applications of artificial intelligence and the design of fixtures for automated manufacturing. IEEE Int. Conf. on Robotics and Automation. 1986:345-351.

24. Prombanpong, S. An optimization approach to the integration of computer aided fixture design with macro process planning for three dimensional prismatic workpiece manufactured on CNC machining centres. Ph.D Dissertation, The Ohio State University, Columbus, Ohio, USA. 1990

25. Boerma, JR and HJJ Kals. FIXES, a system for automatic selection of set-ups and design of fixtures. Annals of the CIRP. 1988; 37(1):443-446

26. Youcef-Toumi, K, JJ Bausch and SJ Blacker. Automated set-up and reconfiguration for modular fixturing. Proceedings of Manufacturing Int. '88, Symp. on Manuf. Sys. - Design, Integration, and Control, ASME, April 17-20. 1988;III:121-128

27. Boerma, JR and HJJ Kals. Fixture design with FIXES: The automatic selection of positioning, clamping and supporting features for prismatic parts. Annals of the CIRP. 1989; 38(1):399-402

28. Ngoi, BKA, K Whybrew and RJ Astley. Computer aided assembly of modular fixturing systems. Proceedings, Asia-Pacific Industrial Automation (IA) '90 Conf on Automation, Singapore. 1990: 593-607

29. Pham, DT and A De Sam Lazaro. Autofix - An expert CAD system for jigs and fixtures. Int. J. Mach. Tools Manufact. 1990;30(3):403-411.

30. Darvishi, AR and KF Gill. Knowledge representation database for the development of a fixture design expert system. Proc. Ins. Mech. Engrs. 1988;202(B1):37-49.

31. Darvishi, AR and KF Gill. Expert system for fixture design. Int. J. of. Prod. Res. 1990;28(10):1901-1920

32. Chan, KC, B Benhabib and MQ Dai. A reconfigurable fixturing system for robotic assembly. J. of Manuf. Sys., 1990;9(3):206-221.

33. Youcef-Toumi, K and JH Buitrago. Design of robot- operated adaptable fixtures", Proceedings of Manufacturing Int. '88, Symp. on Manuf. Sys. - Design, Integration, and Control, ASME, April 17-20. 1988;III:113-119

34. Giusti, F, M Santochi, and G Dini. Robotised assembly of modular fixtures. Annals of the CIRP. 1991; 40(1):17-20

35. Michalski, RS and RE Stepp. Learning from observation: Conceptual clustering. Machine Learning Vol II: An Artificial Intelligence Approach (Chapter 11). Eds: RS Michalski, JG Carbonell and TM Mitchell, Los Altos, CA. 1986:331-363

36. Stepp, RE and RS Michalski. Conceptual clustering: Inventing goal-oriented classification of structured objects. Machine Learning Vol II: An Artificial Intelligence Approach (Chapter 17). Eds: RS Michalski, JG Carbonell and TM Mitchell, Los Altos, CA. 1986:471-498

37. Senthil kumar A. A knowledge-based generative modular fixture design methodology for prismatic workpieces. Ph.D thesis, National University of Singapore, Singapore. 1992

38. Van Houten, FJAM. PART: A computer aided process planning system. University of Twente, Netherlands. 1991

39. Hargroven, A and A Kusiak. Computer-aided fixture design: a review. Int. J. of Production Research. 1994;32(4):733-753

40. Senthil kumar A., AYC Nee and S Prombanpong. Expert fixture-design system for an automated manufacturing environment. Journal of CAD. 1992;24(4):316-326

6 Analysis Methods for Workpiece Distortion and Deflection

6.1 Introduction

During machining operations, various forces are present, viz., cutting forces, inertia forces due to reversal of motions of machine tables, gravitational forces due to the weight of a part and centrifugal forces if it is revolving. In order to counter the aforementioned forces, resisting or clamping forces are applied so that these forces can be properly balanced and maintained by the clamping action so that a part can remain in its correct relation to the locating elements, and hence the cutting tools. Under the complex interaction of the forces present, a part is likely to deform, especially in thin and unsupported sections. Workpiece distortion and deformation would directly affect the dimensional and form accuracies that can be achieved.

Very few fixture designers would specify the magnitude of the clamping force to be applied as its estimation is difficult and errors may occur due to unexpected variations in cutting forces or the coefficient of friction [1]. A rigorous analytical study of the fixture-workpiece system would require the formulation of kinematic and dynamic models involving non-linear algebraic equations of kinematics and non-linear differential equations of dynamics [2]. Even if clamping force magnitudes are appropriately specified by the designers, it will be difficult for the shopfloor operators to follow the specifications unless the clamps are power-operated where pneumatic or hydraulic pressure can be adjusted accurately. Under most situations, shopfloor operators would use their judgement in tightening the clamps. Unfortunately, this may not be done consistently and most of them tend to apply larger than necessary clamping forces just to be "on the safe side" and this causes unnecessarily large deformation and high stresses in the workpiece. An optimal design of the fixturing system has to do with the minimisation of clamping forces and the least workpiece deformation [3].

This chapter first summarises the clamping requirements and examines the various types of clamps commonly used in fixtures and the theoretical/empirical clamping forces which can be achieved. Machining force models based on milling and drilling are reviewed as they are required in estimating clamping forces. Analytical aspects of estimating workpiece distortion is presented with current research work conducted on this aspect.

6.2 Clamping Devices

6.2.1 Basic Requirements

The basic requirements of clamping devices are as follows:

- The clamping action should not disturb the equilibrium position of the workpiece with respect to the locating and supporting elements.
- Present minimum hazards to the operators.
- Fast clamping action, and the magnitude of clamping force can be repeatedly controlled.
- Clamps should not interfere with cutter path, loading and unloading of a workpiece.
- Clamping forces must force the workpiece into contact with all locators (axiom 16)

6.2.2 The Magnitude and Distribution of Clamping Forces

The magnitude of the clamping force must be appropriate: it should be large enough to prevent any movement of the workpiece and yet not too excessive as to cause deformation of workpiece and fixture elements. The correct clamping force that should be applied during a machining operation is dependent on many factors, some of which are listed as follows:

- Workpiece property, *i.e.* whether a workpiece is made of hard alloy or soft material.
- Workpiece geometry in terms of structural rigidity and presence of thin wall sections and easily deformable features.
- Number of clamps to be used in the fixture configuration and how clamping forces are distributed. The sequence in which the clamps are

actually applied to the workpiece is also important. Generally, the major clamp, *i.e.* the one that applies the main clamping force, should be applied last.

- An even force distribution when a clamp is used to hold more than one workpiece or a single clamp providing several clamping points on the same workpiece. In case of size variation, equalisers have to be used.

- Type of machining operation, in particular, variation in the magnitude and direction of machining forces.

- Machining parameters, *i.e.* depth of cut, feedrate, maximum material removal rate, etc. which directly govern the magnitude of cutting forces.

- Type of clamp to be used, and whether clamping action is mainly due to friction alone.

- Tribological surfaces and surfaces where good surface finish must be maintained should have restricted or controlled amount of clamping force.

6.2.3 Types of Clamping Devices, Clamping Action and Theoretical Estimation of Clamping Forces

Clamping devices commonly used in fixturing work can be broadly classified as shown in Figure 6.1. Clamping elements may be manually operated or by means of electrical, pneumatic, hydraulic or other power sources. They can also be classified according to the type of mechanisms used. The two basic classes are: (a) application of the inclined plane theory, *i.e.* screws, wedges, cams, *etc.* and (b) application of lever principle, *i.e.* levers, toggles, *etc.*, or a combination of both, *e.g.* a strap clamp. Some of the more commonly used clamps are described in terms of their clamping actions and theoretical estimation of clamping forces in the following sections.

6.2.3.1 Screw clamps

The action of screw clamps is based on the theory of wedges and is similar in operation compared to wedge and cam clamps. A screw thread can be considered as a wedge wound around a cylinder while a cam is a wedge folded around a circle. Detailed mechanism and analysis of the action of screws can be found in

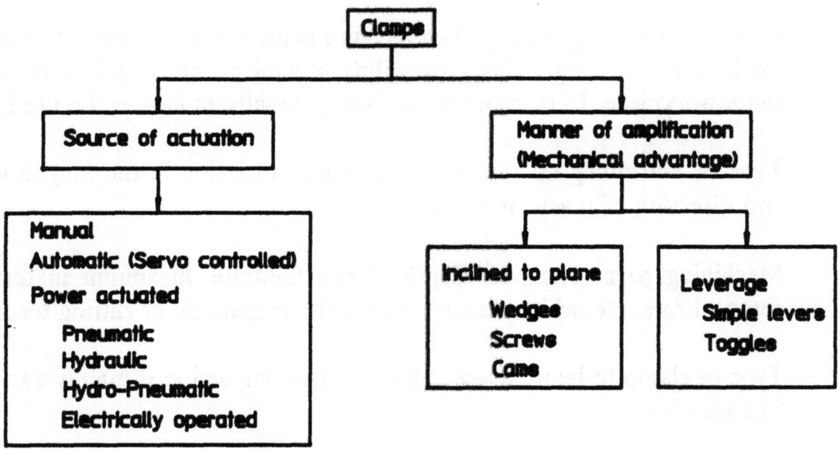

Figure 6.1 Classification of clamping devices

many machine design textbooks. In fixture design, approximate formulae for estimating clamping torque and force relation will suffice.

For a screw of nominal diameter D and an operating coefficient of friction μ, the torque T required to produce a clamping force P is given by the following expressions [4]:

$$T = 0.2DP \qquad (\mu = 0.15)$$
$$T = 0.164DP \qquad (\mu = 0.12)$$
$$T = 0.139DP \qquad (\mu = 0.10)$$
$$T = 0.115DP \qquad (\mu = 0.08)$$

In an average workshop condition, a coefficient of friction $\mu = 0.15$ is considered a representative value.

Tightening of screws is usually done manually and as a result, the clamping force is not uniform and may vary from operator to operator. As the force is usually applied through a lever which may be in the form of a wrench or a knob, the length of the lever or the size of the knob is also important. Average data [4] show that for levers up to 20 cm in length, a 5 N per cm lever length can be taken as the average force for one-hand operations while a torque of 211 N-mm per mm knob diameter can be applied with round and four-lobe hand knobs.

Using the above empirical formulae, for a hand knob of 60 mm diameter, the clamping force with a M10 screw is computed as follows:

$$T = 211 \text{ N-mm/mm} \times 60 \text{ mm} = 12{,}660 \text{ N-mm}$$
$$P = T/0.2D = 12{,}660 \text{ N-mm}/(0.2 \times 10 \text{ mm}) = 6{,}330 \text{ N}$$

6.2.3.2 *Strap clamps*

The family of strap clamps all operate on the same principle, *i.e.* the clamping force is derived from the tightening of screws and it can be magnified or reduced through the principle of levers. The physical arrangement of strap clamps may vary according to one of the several types of levers as shown in Figure 6.2 [4].

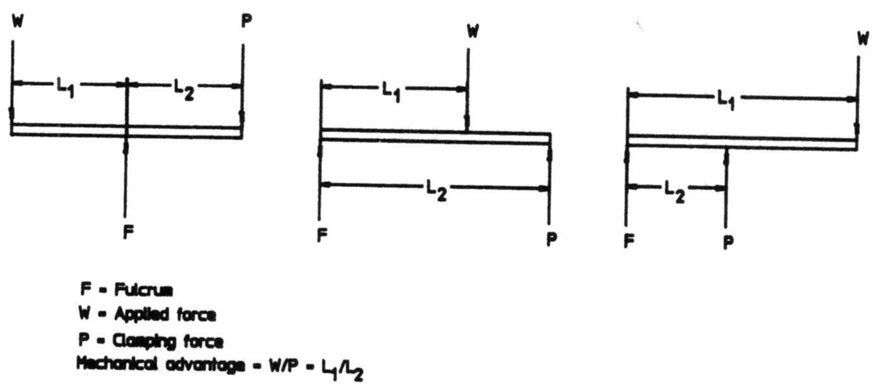

Figure 6.2 The basic classes of levers used in strap clamps

In most of the cases, it is assumed that the magnitude of P, the clamping force is known and it is required to find the applied force F.

For the different cases, the ratios of P to F are as follows:

(a) $P/F = L_1/L_2$
(b) $P/F = 1$
(c) $P/(F_1 + F_2) = 1$
(d) $P/F = L_1/L_2$
(e) $P/F = L_1/L_2$

All the above cases can be either screw, cam or actuated by other means.

6.2.3.3 Toggle clamps

Toggle clamps are essentially similar to the eccentric clamps although the dimensions and the design of the moving parts are largely different. They have a good ratio of holding force to application force and a positive locking action with fast operations.

The toggle mechanism uses an over-centre condition to close and lock its clamping action as shown in Figure 6.3. From the diagram, $R = F/2 \cot a$, where F is the applied force, R is the clamping force and a is the instantaneous angle with the sliding link with the horizontal.

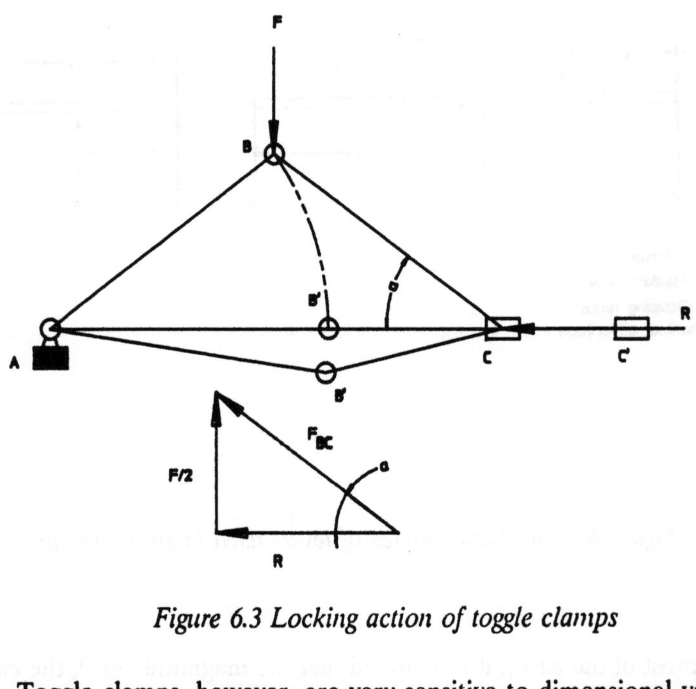

Figure 6.3 Locking action of toggle clamps

Toggle clamps, however, are very sensitive to dimensional variations as the clamping capacity can vary, theoretically, from zero to infinite clamping force, depending on the position of the clamping point. The maximum clamping forces are typically listed in workholding catalogues for toggle clamps. The maximum clamping for an input force occurs at the position when the toggle clamp just goes over-centre before locking. Estimating clamping point at other positions is not an easy task [5].

6.2.3.4 Power clamps

Power clamping provides consistent clamping forces compared to manually operated clamps. They are also fast acting and are often employed in repetitive clamping applications. They are often used in flexible manufacturing systems where machining operations may require certain clamps to be retracted to allow passage of cutting tools. Power clamps are normally operated by hydraulic pressure, pneumatic pressure, or a combination of both.

Clamping forces can be estimated quite easily based on operating pressures (typically from 20 to 35 MPa), effective area of cylinders and mechanical advantages of the clamping mechanism.

6.3 Machining Forces, Force Models for Milling and Drilling

During a machining operation, the cutting tool travels in space and the cutting force may vary both in direction and magnitude. Chou [6] considered the effects of all cutting forces to constitute a "force field", which is a function of both time and space. He computes the envelope of this force field and strategically lays out locating and clamping elements to neutralise the cutting forces.

Both the mechanistic and empirical force models can be used to predict machining forces. The mechanisitc model is one in which mathematical formulae are used to describe the cutting mechanics. The empirical force model, on the other hand, uses experimental data and constants to establish simple empirical equations to extrapolate and/or interpolate the machining forces.

In this chapter, only milling and drilling operations are considered as they are most commonly found in a 3-axis vertical machining centre.

6.3.1 Milling Force Models

The basic milling processes are peripheral (slab) and end (face) milling. In peripheral milling, usually a horizontal milling machine is used and the surface being milled is parallel to the cutter axis. In end milling, the surface is perpendicular to the cutter axis. Only end milling operations are considered here as it is most common in a vertical machining centre.

A mechanistic model allowing for cutter eccentricity developed by Armarego and his associates [7], [8], [9] is adopted for predicting end milling forces. Besides this model, two other models have been developed, viz., the "ideal" cutter model where cutters are assumed to be rigid and without eccentricity and the "deflection" model where the eccentricities as well as the deflection of the cutter during machining are considered. The "eccentricity" model gives good predictions of average machining forces and is relatively easier to incorporate. The "ideal" model gives rather simplisitc results while the "deflection" model, although known to be the most accurate, is rather complex and requires long computational time. The eccentricity model is adopted in the present analysis.

6.3.2 Drilling Force Models

Empirical formulae are used to estimate the drilling thrust and torque as mechanistic models are complex and require certain data before they could be used. Drilling thrust and torque are functions of the drill diameter, drill chisel edge length, feed per revolution and workpiece material.

The following empirical formulae for drilling thrust and torque have been adopted [10].

$$F = 2Kf^{0.8}d^{1.8}B + Kd^{2}E$$
$$T = Kf^{0.8}d^{1.8}A$$

where F is the drilling thrust in N
T is the drilling torque in Nm
f = drilling feed in m/rev
d = drill diameter in m
K = workpiece material constant
A,B,E = drill design constants

6.4 Theoretical Formulation of Workpiece Deformation Using Analytical Methods

In a workpiece-fixture interaction system, the accuracy of a machined workpiece depends to a large extent on its rigidity in terms of proper supporting, locating and clamping considerations. Unsupported portions will deflect and deform elastically under large cutting and clamping forces resulting in undesirable displacement and loss of dimensional accuracy. Local plastic deformation may also occur directly at the clamping and locating points. The correct magnitude of the clamping forces is therefore critical in maintaining workpiece accuracy. In the present approach, clamping forces required at the clamps are derived based on Coulomb's friction models and maximum machining forces and torques which may be experienced. Similar approaches have also been adopted by various other researchers [1], [3], [11].

6.4.1 Modelling of Workpiece Deformation

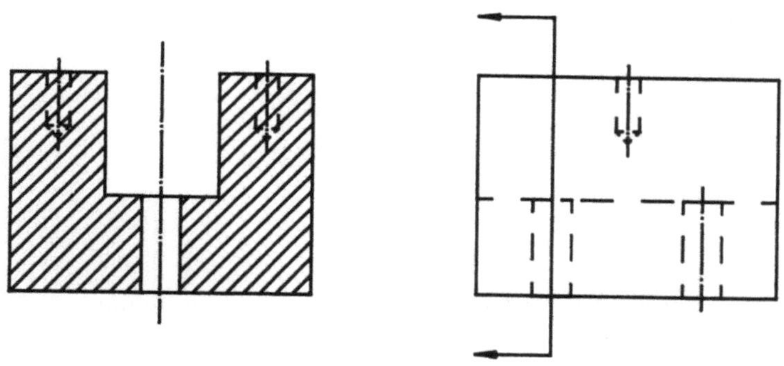

Figure 6.4 A prismatic workpiece to be considered in the deformation analysis

Workpiece deformation under clamping and machining forces is modelled using the finite element method. Figure 6.4 shows a prismatic workpiece where the machining operations involved are a through slot across the workpiece, several small blind holes on the top face and a number of through holes at the bottom of the slot. There are two possible set-up configurations: two set-ups or a single set-up. Figure 6.5 illustrates the two set-up operation as the workpiece has to be raised or inverted to allow the through holes on the bottom of the slot to be drilled. Figure 6.6 shows the arrangement of a single set-up operation as the workpiece has already been elevated on parallel bars prior to milling the slot. The configuration shown in Figure 6.5 is more rigid as the workpiece is placed directly onto the base of the fixture. The effect of milling and clamping forces on the possible deformation of the workpiece is deemed to be smaller in this configuration compared to the one shown in Figure 6.6. The trade-off, however, is between set-up rigidity and possible loss of accuracy due to relocation of the workpiece.

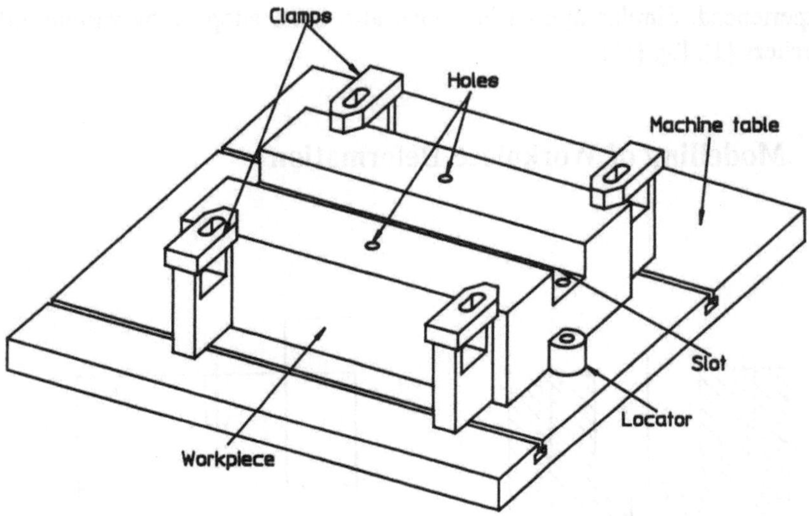

Figure 6.5 Workpiece configuration for two set-ups

A finite element method is used to estimate the workpiece deformation and findings are recommended to the fixture designer whether it is more practical to opt for a two set-up or a single set-up operation.

6.4.2 Parameters Considered in Simulating Workpiece Deformation

The following assumptions are made in simulating workpiece deformation.

(1) The clamping forces are assumed to be uniformly distributed,
(2) The drilling thrust is assumed to be a point load.

Since milling forces can vary both in direction and magnitude, only the largest magnitude and the most critical direction are considered based on an end-mill with a radius of 20 mm, a feedrate of 0.254 mm/tooth and a machining speed of 0.24 m/min. For drilling the 10 mm diameter holes, the feedrate assumed is

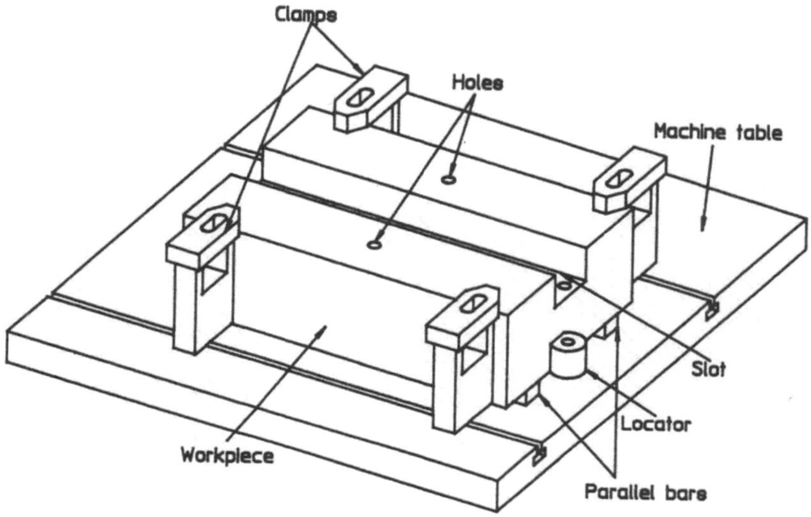

Figure 6.6 Workpiece configuration for a single set-up

0.30 m/min. The milling forces, drilling torque and thrust are calculated using milling and drilling force models outlined in Sections 6.3.1 and 6.3.2.

Workpiece deformation is affected by the following variables.

(1) The height and breadth of the workpiece,
(2) The position of the clamps and the parallel bars,
(3) The size of the parallel bars.

The depth of the workpiece has not been considered in this analysis, as the finite element method used is essentially two-dimensional. The critical section analysed corresponds to the position where maximum milling forces would occur. The machining parameters and the size of the parallel bars are assumed to remain constant.

Four workpiece proportions with various slot depths have been used in simulating workpiece deformation (see Figure 6.7). The other variables considered in the simulation are: length of the clamps (L) over the workpiece, the position of the parallel bars from the edge of the workpiece (D) and the depth of the slot from 5 to 40 mm in increments of 5 mm. The width (W) of the parallel bars is kept constant at 20 mm. A typical mesh size used is 272 elements with 316 nodes as shown in Figure 6.8.

The values of L and D for each workpiece are summarised in Table 1.

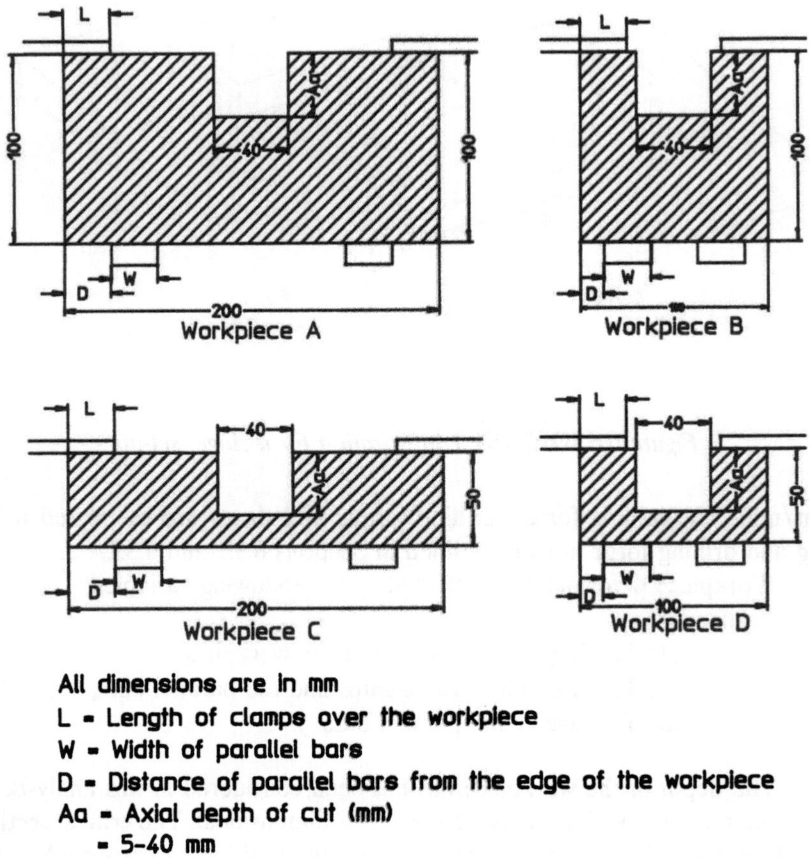

Figure 6.7 *Various workpiece proportions used to estimate workpiece deformation*

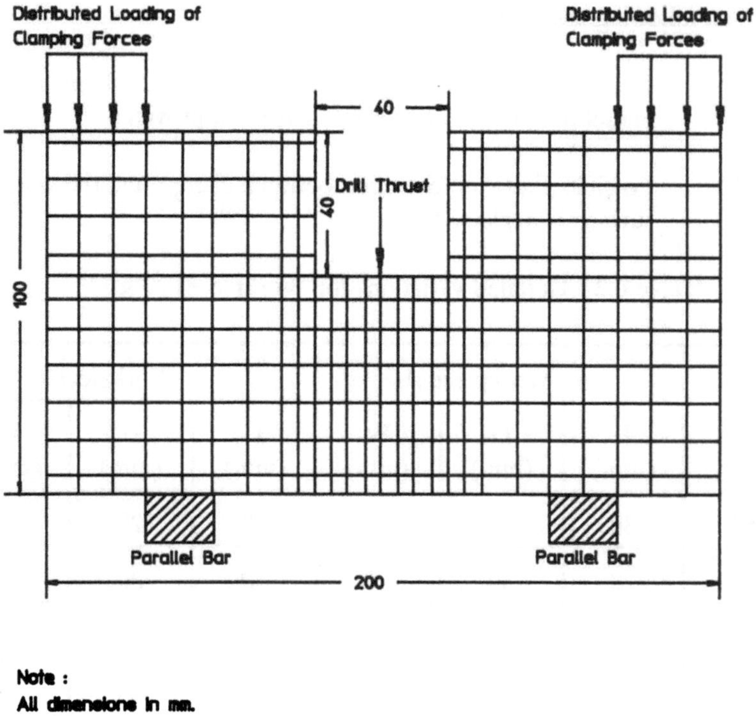

Figure 6.8 Finite element mesh of a workpiece

Table 6.1 Values of L and W for each type of workpiece

	L (mm)	D (mm)
Workpiece A	30	30
Workpiece B	15	15
Workpiece C	30	30
Workpiece D	15	15

For workpieces A and C, the width (W) of the parallel bar is further varied as shown in Table 2.

The following assumptions are made:

(1) Clamping forces are estimated based on Coulomb's friction model and no factor of safety has been provided.

(2) Possible plastic deformation between clamps and the workpiece surface is neglected.

Table 6.2 Detailed variations for Workpieces A & C

	Workpiece A			Workpiece C		
	L (mm)	D (mm)	W (mm)	L (mm)	D (mm)	W (mm)
Config I	30	30	20	30	20	30
Config II	20	30	20	30	30	30
Config III	30	10	35	30	60	30
Config IV	50	20	30	30	30	25

6.4.3 Simulation Results

6.4.3.1 Using parameters as shown in Table 1

The workpiece deflection points that are of particular interest are points A and B as shown in Figure 6.9. Point A reflects the shape of the slot as any tapering or opening will be reflected by the movements of this point. Point B represents the flatness of the base of the workpiece. If the workpiece is required to sit properly, flatness is of critical importance. Exaggerated deformations of workpieces A&D based on finite element analysis are shown in Figure 6.10.

Figure 6.11 shows the deflection at Point A for four different workpieces at various axial depths of cuts. Workpiece A has the least deflection as it is structurally more rigid compared to the other workpieces. Workpiece C experiences a deflection in the opposite direction, i.e. Point A moves inwards or the slot narrows at the opening. The deflection is due to an interaction of the forces arising from the drilling and milling operations as well as the clamping action. For workpieces B and D, the slot opens up, due mainly to the clamping action.

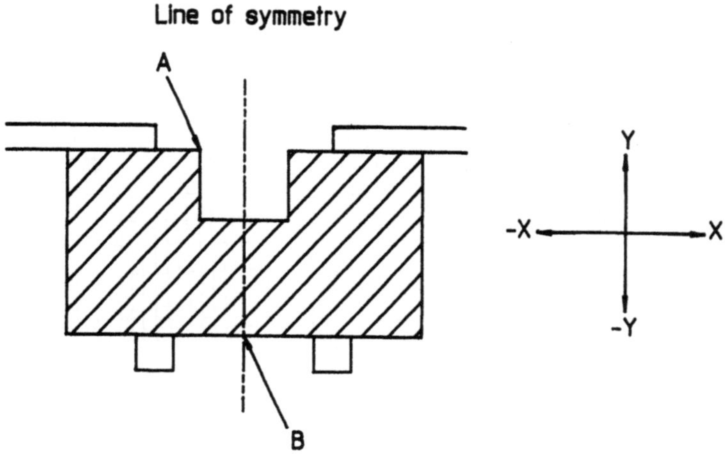

Figure 6.9 Deflection points of interest

Figure 6.12 shows the deflection at Point B for the same parameters considered. Workpiece B has the lowest deflection as the parallel bars are placed at a closer span and this has provided better support.

6.4.3.2 Using parameters as shown in Table 2

Only Workpiece A results are discussed as somewhat similar results have been obtained for Workpiece C.

Figure 6.13 shows the deflection at Point A for Workpiece A. Noted that the vertical scale has been given a ten-fold increase as the deflection obtained is quite small. Configuration I is the same curve for Workpiece A as shown in Figure 6.11. In general, larger parallel bars give better support and lower deflection.

Similar conclusions can be drawn. Figure 6.14 shows the deflection at Point B for Workpiece A.

6.5 Conclusions

Workpiece deflection simulation enables a process planner to optimise the clamping arrangements for any particular workpiece configurations. By changing

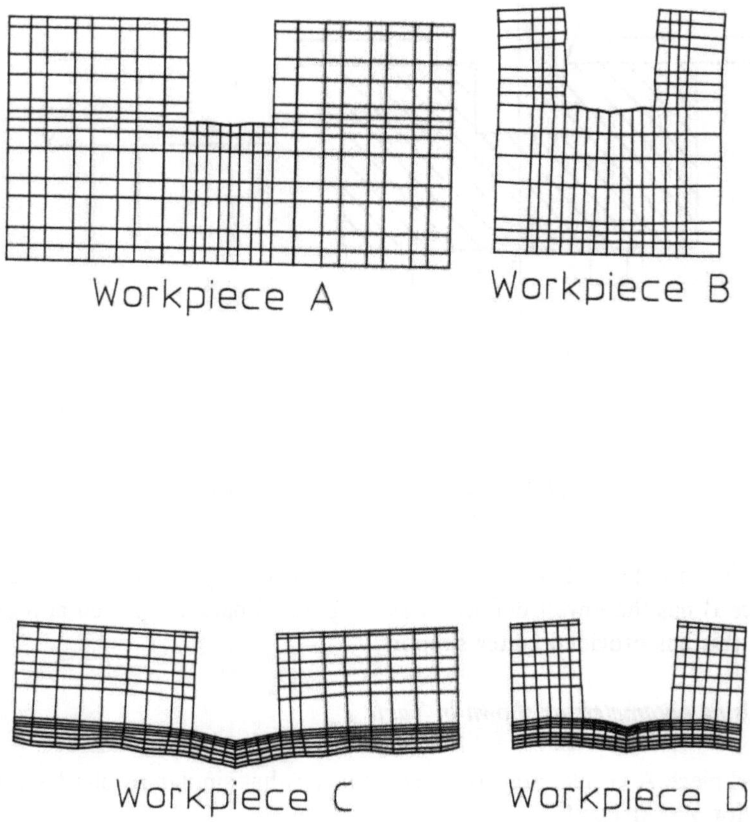

Figure 6.10 Exaggerated deformation of the workpiece using FEA

the lengths of the clamps and the widths and positions of the parallel bars in the simulation models, the process planner can examine the resulting effects of the deflections of the workpiece at any critical points. In this particular example, a single set-up is preferred as analyses show that the workpiece deformation is quite small and has negligible effects on workpiece tolerances. A single set-up would also increase accuracy and save resetting time needed. With a fast computer, the process planner would be able to arrive at a particular fixturing configuration

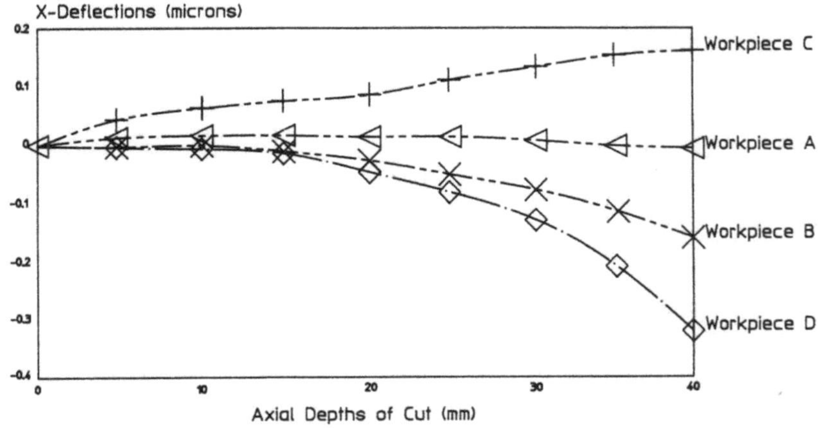

L = 30 mm for workpiece A,C; L = 15 mm for workpiece B,D; W = 20 for all workpieces

Figure 6.11 Deflections (microns) at Point A vs axial depth of cut

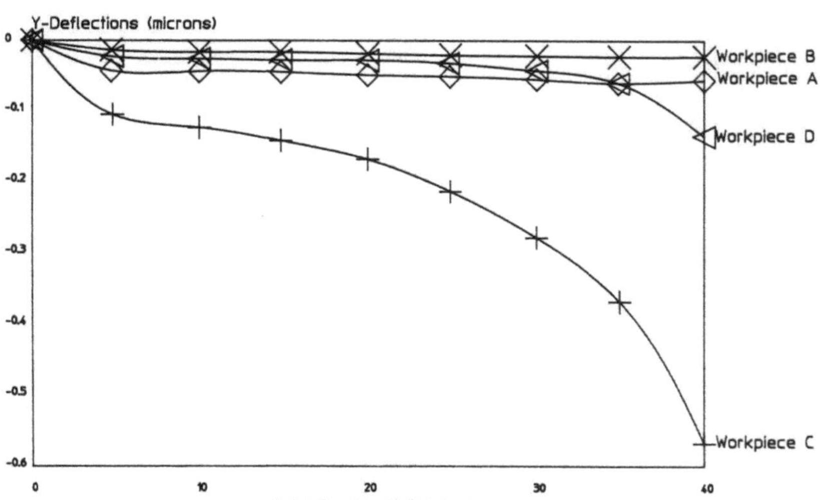

L = 30 mm for workpiece A,C; L = 15 mm for workpiece B,D; W = 20 for all workpieces

L = 30 mm for workpiece A,C; L = 15 mm for workpiece B,D; W = 20 for all workpieces

Figure 6.12 Deflections (microns) at Point B vs axial depth of cut

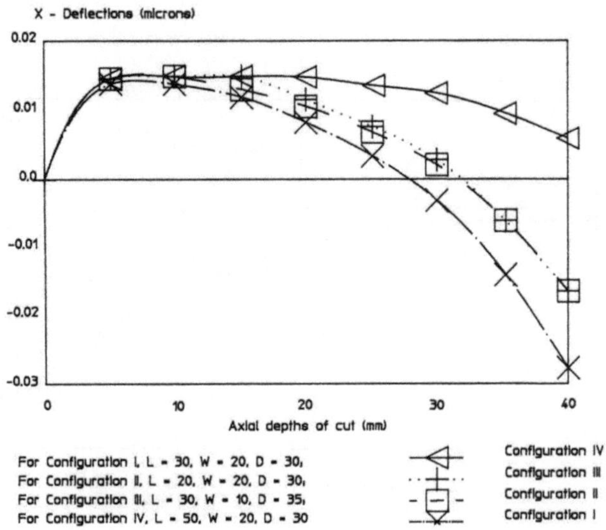

Figure 6.13 Deflection (microns) at Point A considering the size of the support

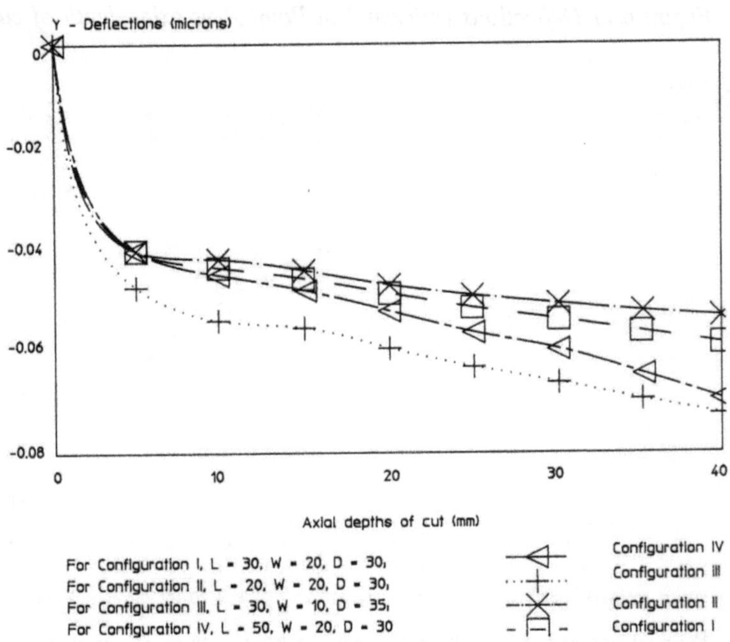

Figure 6.14 Deflection (microns) at Point B considering the size of the support

within a relatively short time. The accuracy of the analyses, however, depends on the quality of the force models as well as the evaluation of the clamping forces required.

6.6 References

1. Lee, SH and M Cutkosky. Fixture planning with friction. 1989 NSF Design Theory and Methodology Conference, Amherst, MA. June, 1989:1-28

2. Mittal, RO, PH Cohen and BJ Gilmore. Dynamic modeling of the fixture-workpiece system. Robotics & Computer-Integrated Manufacturing. 1991;8(4): 201-217

3. Lee, JD and LS Haynes. Finite-element analysis of flexible fixturing system. Trans. of the ASME Journal of Engineering for Industry, May 1987; 109:134-139

4. Henriksen, EK. Jig and Fixture Design Manual. Industrial Press Inc. 1973

5. Noaker, PM. Commonsense clamping. Manufacturing Engineering, October 1992;43-47

6. Chou, Y-C, V Chandru and MM Barash. A mathematical approach to automatic configuration of machining fixtures: analysis and synthesis. Trans. of the ASME Journal of Engineering for Industry, November 1989;111:299-306

7. Armarego, EJA and RC Whitfield. Computer based modelling of popular machining operations for force and power prediction. Annals of CIRP. 34(1):25-29

8. Armarego, EJA and NP Desphande. Computerised predictive cutting models for forces in end milling including eccentricity effects. Annals of CIRP. 38(1):45-49.

9. Armarego, EJA and NP Desphande. Computerised end-milling force predictions with cutting models allowing for eccentricity and cutter deflections. Annals of CIRP. 39(1):25-29

10. Society of Manufacturing Engineers. Tool and Manufacturing Engineering Handbook. 1988;1

11. Sakurai, H. Automatic set-up planning and fixture design for machining. Journal of Manufacturing Systems. 11(1):30-37

7 Future Trends and Developments in Fixturing Methodologies

7.1 Introduction

Advances in the technology of flexible manufacturing systems, and the introduction of concurrent engineering of product and process design have propounded the need for computer aided fixture design systems. Computer aided process planning is the interface between design and manufacturing and has a key role in achieving integration between the various modules in an FMS environment. Fixture design is inseparable from process planning, and hence to be of practical use any CAPP system must include fixture design. The past few years have seen great advances in automated fixture design systems and the tools used to achieve them. With the exception of fixture design for a limited range of products with close family similarity, fully automated design of practical fixtures is not yet achievable. The advances described in this book have widened the scope of application of computer aided fixture design and provide enhanced links with other product and process design activities.

In this Chapter we will attempt to predict new directions for research in fixture design and future trends in the development of fixtures.

7.2 An Industrial Perspective on Fixture Design

The situation today

In recent years industrial competitiveness has been maintained by increasing productivity. This productivity increase has been made possible by automation of routine design and manufacturing tasks using computers, numerical control of manufacturing processes and robots for materials handling. However, almost 85% of all process plans and fixture designs are still created manually, and detailed optimized plans are rarely produced. Even when companies do use a CAPP system, it is generally done in isolation from product and fixture design. This clearly indicates a lack of confidence in commercially available CAPP systems. Alting and Zhang, in a comprehensive review of 156 CAPP systems, published in 1989, came to the conclusion that "....the research and implementation of CAPP are not matching industrial expectations and satisfactory results are not provided by existing systems for real production environments." [1,page 571]. It is our opinion that there has been little improvement in this state of affairs in the five years since this paper was published.

Fixtures represent a considerable investment, some estimates put it as high as 40 to 60 % of the total production cost. Despite the obvious importance of optimising such a significant part of production costs, in most industries, the design and assembly of fixtures has not benefited from computer aids and is still the province of skilled craftsmen.

Careful assessment of this situation and the underlying factors is essential to elucidate discussion on the direction of future research and development. The following are some of the observations made in industries using either fixture design or CAPP systems:

- Fixture design is not an independent task. Current systems offer only partial solutions and do not provide good interfacing among process planning, scheduling, tool management and product design as well as existing databases.

- Tools which focus on fixture design synthesis rather than analysis are needed to help designers make decisions regarding fixturing techniques.

- Most industries still use dedicated fixtures instead of applying newly emerging technologies such as flexible and modular fixtures.

- Users require faster data processing systems.

- Systems should be open and flexible and should not have black-boxes containing heuristics and algorithms which cannot be altered.

- Systems should be function-orientated so that they can be assimilated by industry quickly.

- Systems should be flexible enough to accommodate a wide range of products.

- Easy access to knowledge bases, databases and catalogues is needed while designing a fixture.

Situations and trends in the next decade

- Rising wages and increasing demands for improvements in working environments will force industry to increase the degree of automation and mechanization.

- Manual assembly and design of fixtures consume a considerable amount of time and a dramatic increase in development and commercialization of automated fixture design systems is expected.

- Robots and vision systems will gain popularity both for building fixtures and orienting workpieces for machining.

- Modular fixtures will be used widely because of their advantages over dedicated fixtures.

- The development of new fixtures and design methodologies to a large extent depends on process developments and production methods. Automated fixturing systems are likely to cover not only machining fixtures but also the following:

  - Micro-machining fixtures
  - Welding fixtures
  - Inspection fixtures
  - Heat-treatment fixtures
  - Assembly and dis-assembly fixtures

7.3 Future Design Methodologies

The development of fixture design methodologies to suit rapidly changing design technologies is discussed in this section.

The emphasis during the next decade will be on concurrent engineering of product and manufacturing system design and the fixture designer will have to interact with the various manufacturing modules in an FMS environment. Three different fixture design methodologies to meet the challenges of the rapidly changing environment are discussed below:

- Concurrent engineering and fixture design
- Generic fixture design systems
- Case-based learning

7.3.1 Concurrent Engineering and Fixture Design

Concurrent engineering and the introduction of life-cycle considerations in FMS necessitate a two-way interaction between the design of a product and its fixtures. It is no longer sufficient to ensure an effective one-directional flow of information from product design to fixture design. Feedback from the fixture designer is essential to assist the product designer to assess alternative design features at an early stage, not only with regard to function, but also manufacturability, assemblability, cost. A large percentage of the cost of a design is committed once its features, material, tolerances and surface quality parameters are selected. However, if early in the design process, a fixture designer realizes that the designed feature requires expensive fixturing, he/she can feedback this information to the product designer and design changes can be made.

Fixturability, or the ease of fixturing a part, is very much a manufacturability concern. However, this activity has not received sufficient attention in the past. Nee and Senthil kumar [2] have proposed using an expert system to assess the fixturability of a product design by considering the features present in the product model. They are classified as clamping, locating and supporting features.

The features of a product model are classified in general as strong and weak features. Using the knowledge embedded in the expert system, weak and strong features are identified by comparing the size, approachability of the feature, unsupported features *etc.* This information can be used to identify the machining sequence of the product model. In general the fixturability of a workpiece can be evaluated by examining the following feature characteristics [2]:

- Size for locating, clamping *etc.*
- Tolerance relationships
- Machining precedence relationships
- Accessibility
- Force transmissibility
- Symmetry

The process of evaluating the fixturability of a workpiece is as follows:

Step 1: Group all the features by considering machining precedence and tolerance relationships into individual partial set-ups.
Step 2: Check size and accessibility of features.
Step 3: Assess force transmissibility and feature symmetry.

By examining the fixturability of a workpiece, it is to some extent possible to determine its manufacturability. Weak features could be improved during the design stage if these changes do not interfere with their intended functions.

7.3.2 Generic Fixture Design

In general, manufacturing activities start with process planning coupled with the design of fixtures. A fixture designer initially considers the process sequence and the features to decide the appropriate type of fixturing elements. This is useful in identifying the initial feasibility of the fixtures.

Current fixture design systems are designed for a particular machining operation such as milling, drilling [3], turning, etc. However producing a part may require various combinations of machining operations. Hence it is convenient to have a generic system which has the ability to adopt the appropriate fixturing knowledge based on the different machining operations involved. This may help to determine a better machining sequence and reduce knowledge duplication.

Generative fixture design is concerned with the initial choice of the manufacturing activity, features and their machining sequence. This information assists the product designer in choosing the appropriate design features at an early stage. Future generative fixture design systems, will be coupled with a solid modeller and an expert system. The design rules, which are usually tailored for a particular manufacturing activity, will be embedded in the expert system which can be used to make inferences about the appropriate fixture elements to be used. Generic fixture design systems, on the other hand, will consist of an expert system where the rules are embedded in the form of conditions or constraints. This can be linked with the individual features such that if a particular feature exists then a particular method of clamping or locating is suggested based on the corresponding constraints. In this manner, both information on the product and the method of fixturing will be available.

This approach will provide the designer with expert advice on aspects of fixturing during the early stages of product modelling. Once the method of fixturing and the features are decided, further detailed design will be carried out using the generative fixture design system.

7.3.3 Case-Based Learning

Case-based reasoning techniques as explained in Chapter 5 will be useful to retrieve existing fixture designs from the database and modify them to suit new workpieces. Machine learning techniques, such as inductive learning, can be coupled with the case-based system to develop systems that can design fixtures automatically. Such intelligent systems will have a data pre-processor, rule induction, a knowledge-based system and human-computer interface. The raw data from previous designs can be processed in the data pre-processor and noise such as illegal attribute values and missing values identified, thus enabling easy knowledge acquisition. The acquired data can then be processed to deduce the production rules. This can be fed into the knowledge-based system for future designs. The knowledge-based system will be automatically reconstructed and updated whenever a new case or new concepts are encountered. The development of rule-based systems using the case-based learning system will help in automating the knowledge acquisition process, assist in the knowledge-base construction and also reduce the problems involved in maintaining the knowledge-base of the system.

7.4 Fixturing Techniques

Flexible fixtures ideally should be single devices capable of holding a family of parts of various shapes and sizes as they are subjected to a wide variety of external forces and torque fields. Attempts to satisfy this ideal have focussed efforts on solving the generic problem of flexible fixturing to enhance productivity, flexibility and cost effectiveness in an FMS environment. Principal flexible fixturing methodologies such as modular fixturing, phase change fixturing and related research have been discussed in Chapters 4 and 5 respectively. This section will discuss the prospects of modular fixtures and fluidized bed fixtures.

7.4.1 Modular Fixtures

Research in the area of modular fixtures has predominantly concentrated on the development of automated modular fixture design systems. Commercially available modular fixtures are intended for general machining processes such as milling, drilling and turning, *etc* and the automated fixture design systems developed so far have been restricted to designing fixtures for these operations. However, future machining processes are expected to concentrate more on micro-machining and micro-grinding. Material increase technologies such as vapour deposition and stereolithography are also gaining popularity. Eventually, we can expect to see fully automated formative processes, such as auto-configurable moulds and electromagnetic forming. Another trend will be more versatile combinations of processes in single machines, much like today's CNC machining centres, but incorporating additive, subtractive and forming processes. Hence there will be a need to develop fixtures which satisfy the requirements of these new processes. Research will be required to develop hydraulically or pneumatically activated modular elements equipped with force sensors and on-line force regulators to minimize workpiece deformation.

In future, most machining industries will move towards near-net shape designs which may involve light machining with reduced machining forces. Present day modular element systems which are either hole or tee-slot-based are designed to take heavy loads and usually take a few hours to build manually. For practical purposes such systems are unsuited to assembly using robots. With reduced machining forces new fixturing systems with magnetized bases can be used. These will be easier to assemble using robots and it will become possible to rapidly build new fixtures automatically. Research is required in such areas to increase productivity.

Other areas which need increased emphasis in developing both software and hardware (modular elements) are:

- Automated design of micro fixtures
- Automated design of welding fixtures
- Automated design of inspection fixtures
- Automated design of heat-treatment fixtures
- Automated design of assembly fixtures

7.4.2 Phase-Change Fixtures

Phase-change fixturing methodology makes use of the phase-change of certain materials from liquid to solid and back to liquid. Ideally such fixturing methodology is able to accommodate parts of various shapes and geometries. A primary drawback of this fixture type is the difficulty of finding materials that meet the necessary requirements. Pseudo phase-change fixtures are becoming increasingly attractive because of their dual-phase nature. Such fixtures can be used in inspection, welding, assembly, painting and other operations that do not encounter high dynamic loads. Results from various researchers indicate that the increase in compaction pressure and/or workpiece submergence depth increases fixture rigidity. Research is required to establish standards based on material property, compaction pressure and machining load which are critical for the design, analysis and use of particulate fluidized bed fixtures in future industrial applications.

7.4.3 Vision Systems

Vision systems are usually employed in robots for guidance in performing certain tasks. Such vision systems are gaining popularity in FMS for guiding robots to pick-and-place workpieces. These systems are used for automated assembly and disassembly of parts. They can also be used for building modular fixtures, orientating, loading and unloading the workpiece to the fixture, and also in holding the workpiece in fluidized bed fixtures. Research in vision and robot-assisted automated fixture building systems will gain popularity in the next decade.

7.4.4 Dynamic Fixtures to Meet Uncertainties and Changes in Scheduling

The FMS environment consists of high level data processing units, assembly cells, industrial robots, inspection machines together with computer-integrated material handling and storage systems. The most important criterion in such systems is their ability to cope with changing characteristics of products, batch size and mixed orders with maximum utilization of the equipment. In other words the system must be truly flexible. In such an environment the re-routing of processes is possible in the event of machine breakdown. In order to meet the challenges of such an environment the need for dynamic fixtures arises. Dynamic fixtures are fixtures which have the ability to cope with changes in the process plans. Such fixtures may take the form of CNC fixtures or reconfigurable fixtures using mechatronic devices. They would be able to meet the uncertainties and rapid changes in scheduling and product design.

7.5 References

1 Alting L, Zhang H. Computer aided process planning: the state of the art survey. Int. J. Prod. Res. 1989; 27(4):553-585

2. Nee AYC and A Senthil kumar. Assessing fixturability for machining operations on prismatic workpieces. Int. conference on experimental and theoretical mechanics, Dec.7-9, Bandung, Indonesia. 1993

3. Senthil kumar, A, AYC Nee and S Prombanpong. Expert fixture design system for an automated manufacturing environment. Computer Aided Design. 1992;24(6): 316-326

Bibliography

This bibliography is classified into seven sections:

- General background
- Generative and variant fixture design
- Expert systems in fixture design
- Feature recognition and feature-based design for product model
- Modular, phase-change and reconfigurable fixtures
- Tolerance control
- Fixture design analysis

General background

Bidanda, B and CK Muralikrishnan. Flexible fixturing for intelligent Manufacturing. Intelligent Design and Manufacturing, Edited by Andrew Kusiak, John Wiley & Sons Inc, 1992.

Brown, PF and CR Mclean, "Interactive process planning in AMRF", Symposium on Knowledge Based Expert Systems for Manufacturing, The winter annual meeting of ASME, Anaheim, California, Dec. 1986.

Blüco Technik. Modular Fixturing System-Parts Catalog 8/89, 1989

Boyes WE (ed). Handbook of jig and fixture design, 2nd edn. Society of Manufacturing Engineers, Dearborn, Mich, 1989

Burbidge JL. The introduction of group technology. Heinemann , London, 1975

Chang, T.C., "Expert process planning for manufacturing", Addision-Wesley publishing company, USA, 1990.

Cohen, PR and E Feigenbaum. The Handbook of Artificial Intelligence, William Kaufmann Inc., Los Aitos, California, USA. 1982.

Coyne RD, Rosenman MA, Radford AD, Balachandran M, Gero JS. Knowledge-based design systems. Addison-Wesley, Reading Mass., 1990

de Garmo EP. Engineering economics, 4th edn. Macmillan, New York,1967

Drozda TJ, Wicks C (ed). Tool and manufacturing engineering handbook, vol 1. Society of Manufacturing Engineers, Dearborn, Mich, 1988.

Eary DF.,Johnson GE. Process engineering for manufacturing. Prentice-Hall, Englewood Cliffs, NJ, 1962

Groover MP. Automation production systems and computer integrated manufacturing. Prentice- Hall, Englewood Cliffs, NJ, 1987

Groover, MP and EWJR Zimmers. CAD/CAM: Computer-Aided Design and Manufacturing. Prentice-Hall Inc., New Jersey, USA. 1984.

Henriksen EK. Jig and fixture design manual. Industrial Press New York 1973

Hoffman GE (ed). Fundamentals of tool design, 2nd edn. Society of Manufacturing Engineers, Dearborn, Mich, 1984

Institution of Production Engineers. Data memoranda on jig and fixture design. The Institution of Production Engineers, London,1973

Jones EJH, rev Town HC. Production engineering- Jig and tool design 8th edn. Newnes-Butterworths, London, 1972

Jones FD, Ed. Jig and fixture design 4th edn. Industrial Press, New york, 1955

Joshi, P.H., Jigs and Fixtures, Tata-McGraw Hill Pub. Com., New Delhi, India, 1988.

Kochan D., editor, CAM-Developments in Computer-Integrated Manufacturing, Springer-Verlag, 1986.

Korsakov, V, Fundamentals of fixture design, Mir Publishers, Moscow, 1989.

Lu SCY. Integrated knowledge, geometry, and data processing technologies to support concurrent product development practice. Proc of the Int Conf on Computer Integrated Manufacturing 1991; 181-184, World Scientific,1991

Michalski, RS and RE Stepp. Learning from observation: Conceptual clustering. Machine Learning Vol II: An Artificial Intelligence Approach (Chapter 11). Eds: RS Michalski, JG Carbonell and TM Mitchell, Los Altos, CA. 1986:331-363

Nevins DJ, Whitney DE. Concurrent design of products and processes. McGraw-Hill, New York,1989

Nixon F. Managing to achieve quality and reliability. McGraw-Hill, Maidenhead, 1971

Noaker, PM. Commonsense clamping. Manufacturing Engineering, October 1992;43-47

Ohwovoriole MS, Roth B. An extension of screw theory. J of Mechanical Design, Trans of ASME 1981; 103:725-735

Paul, R., "The design and operation of FMS-Flexible Manufacturing Systems", IFS Ltd and North-Holland publishing company, UK, 1983.

Pham, D.T., M.J. Nategh and A. De Sam Lazaro, "Jig and fixture designer's assistant", Int. J. Adv. Manuf. Tech., Vol.4, pp.26-45, 1989.

Sedik H. Jigs and fixtures for limited production. Society of Manufacturing Engineers, Dearborn, Mich. 1970

Society of Manufacturing Engineers. Fundamentals of tool design 3rd edn. Society of Manufacturing Engineers, Dearborn, Mich. 1991

Stepp, RE and RS Michalski. Conceptual clustering: Inventing goal-oriented classification of structured objects. Machine Learning Vol II: An Artificial Intelligence Approach (Chapter 17). Eds: RS Michalski, JG Carbonell and TM Mitchell, Los Altos, CA. 1986:471-498

Thompson, BS. Flexible fixturing-Current frontier in the evolution of flexible manufacturing cells. ASME paper no. 84-WA/Prod-16, 1984.

Van Houten, F.J.A.M., "PART: A computer aided process planning system", University of Twente, Netherlands, 1991.

Williams DJ. Manufacturing systems an introduction to the technologies. Open University Press, 1988

Wilson, F.W., Editor, Tool Engineers Handbook-section 91, 2nd Ed., McGraw- Hill Book Company, 1959.

Generative and variant fixture design

Bausch, JJ and K Youcef-Toumi. Computer planning methods for automated fixture layout synthesis. The American Society of Mechanical engineers, Proc. of Manufacturing International' 90, Vol, I- Advances in materials and automation, Atlanta. March 1990:225-232

Boerma, JR and HJJ Kals. FIXES, a system for automatic selection of set-ups and design of fixtures. Annals of the CIRP. 1988; 37(1):443-446

Boerma, JR and HJJ Kals. Fixture design with FIXES: The automatic selection of positioning, clamping and supporting features for prismatic parts. Annals of the CIRP. 1989; 38(1):399-402

Bopaya Monapa, B., "An integrated CAD-CAM approach for the selection of workholding devices for concentric, rotational components", PhD Dissertation, The Pennsylvania State University, USA, 1987.

Cadabaj, J. Theory of computer aided fixture design. Computers in Industry. 1990; 15:141-147

Choi, B.K., CAD/CAM compatible, tool-oriented process planning for machining centers. Ph.D Dissertation, Purdue University, USA, Dec. 1982.

Chou YC. Automatic design of fixtures for machining processes. PhD Thesis, Purdue University,1988

Chou, YC. Computerized fixture design from solid models of workpieces. Master Thesis, School of Industrial Engineering, Purdue University, West Lafayette, Indiana, USA, 1985.

Chou, Y.C. and M.M. Barash, Computerized fixture design from solid models of workpieces. PED-Vol 21, Integrated and Intelligent Manufacturing, (eds. Liu C.R. and Chang TC) ASME, WAM, Calif, pp. 133-142, Dec 7-12, 1986.

Colbert, JL, R Menassa and WR DeVries. A modular fixture for prismatic parts in FMS. Proc.of North American Manufacturing Research Conference, 1986.

Dong, X, WR DeVries and MJ Wozny. Feature-based reasoning in fixture design. Annals of the CIRP. 1991;40(1):111-114

Gandhi, MV and BS Thompson. Automated design of modular fixtures for flexible manufacturing systems. J of Manu. systems. 1986;5(4):243-252

Grippo, P.M., M.V. Gandhi and B.S. Thompson, "The computer aided design of modular fixturing systems", The Int. J. of Adv. Manuf. Tech., Vol.2, pp.75-88, 1987.

Joneja, A and TC Chang. A generalized framework for automatic planning of fixture configuration. Anjanappan, M, and Anand. D.K., (eds), Advances in manufacturing systems engineering PED Vol.37, The ASME winter annual meeting, Dec.10-15, pp.17-28, 1989.

Markus, A., "Strategies for the automated generation of modular fixtures", Proceedings of Manufacturing Int. '88, Symp. on. Manuf. Sys. - Design, Integration, and Control, Vol. III, ASME, April 17-20, pp.97-103, 1988.

Mittal, RO, PH Cohen, and BJ Gilmore. Dynamic modeling of fixture-workpiece system. Robotics and Comp. Integrated Manufacturing. 1991; 8(4):201-217

Nee, AYC and A Senthil kumar. A frame work for and Object/Rule-based automated fixture design system, Annals of the CIRP. 1991; 40(1):147-151

Nee, AYC, A Senthil kumar, S Prombanpong, and KY Puah. A feature-based classification scheme for fixtures. Annals of the CIRP. 1992; 41(1) :189-193.

Ngoi NK, Whybrew K. A fast spatial representation method (applied to fixture design), International Journal of Advanced Manufacturing Technology 1993; 8(2):71-77

Ngoi BKA, K Whybrew and RJ Astley. Computer aided assembly of modular fixturing systems. Proceedings, Asia-Pacific Industrial Automation (IA) '90 Conf on Automation, Singapore. 1990: 593-607

Pham DT and A De Sam Lazaro. AUTOFIX - An expert CAD system for jigs and fixtures. Int J of Mach. Tools Manufact. 1990;30(3):403-411

Prombanpong, S. An optimization approach to the integration of computer aided fixture design with macro process planning for three dimensional prismatic workpiece manufactured on CNC machining centers. Ph.D Dissertation, The Ohio State University, Columbus, Ohio, USA. 1990

Puah, KY and AYC Nee. A microcomputer-based modular fixturing system. Proc., Asia-Pacific Industrial Automation (IA) '90 Conference on Automation: Strategies for Manufacturing Growth, Singapore. May, 1990:580-592

Sakurai, H. Automatic setup planning and fixture design for machining. Journal of Manufacturing Systems. 11(1):30-37

Senthil kumar A. A knowledge-based generative modular fixture design methodology for prismatic workpieces. Ph.D thesis, National University of Singapore, Singapore. 1992

Senthil kumar A, AYC Nee and S Prombanpong. Expert fixture design system for an automated manufacturing environment. J of Computer Aided Design. 1992;24(6):316-326

Sermsuti-anuwat Y. Computer-aided process planning and fixture design (CAPPFD). PhD thesis, University of Canterbury, 1992.

Tung, K.H., A.N. Poo and A.Y.C. Nee, "A modular fixture configuration design system using a rule/object based expert system approach", Proceedings, Asia-Pacific Industrial Automation (IA) '90 Conf on Automation, Singapore, pp.568-579, 1990.

Tung, K.H., "A modular fixture configuration design system using a rule/object based expert system approach", M.Engg Thesis, Dept. of Mechanical and Production Engg., National University of Singapore, Singapore, 1990.

Youcef-Toumi, K., J.J. Bausch and S.J. Blacker, "Automated setup and reconfiguration for modular fixturing", Proceedings of Manufacturing Int. '88, Symp. on Manuf. Sys. - Design, Integration, and Control, Vol. III, ASME, April 17-20, pp.121-128, 1988.

Expert systems in fixture design

Bidanda B and PH Cohen. Development of a computer aided fixture selection system for concentric, rotational parts. Advances in Integrated Product Design and Manufacturing, Winter Annual Meeting of ASME. PED-Vol.47. November 1990:219-226

Chou, YC and MM Barash. Automatic configuration of machining fixtures: Object constraint reasoning. The American Society of Mechanical engineers, Proc. of Manufacturing International' 90, Vol, IV- Advances in materials and automation, Atlanta. March 1990:15-20

Darvishi, AR and KF Gill, "Knowledge representation database for the development of a fixture design expert system", Proc. Ins. Mech. Engrs., 1988;202(B1) :37-49

Darvishi, AR and KF Gill. Expert system for fixture design. Int. J. of. Prod. Res. 1990;28(10):1901-1920

Englert, PJ and PK Wright. Application of artificial intelligence and the design of fixtures for automated manufacturing. IEEE Int. conf. on robotics and automation, 1986:345-351.

Ferreira, PM, B Kochar, CR Liu and V Chandru. AIFIX: An expert system approach to fixture design. ASME Computer Aided/Intelligent Process Planning. 1986; PED-Vol.19:73-82

Grippo, PM, MV Gandhi and BS Thompson. The computer aided design of modular fixturing systems. The Int. J. of Adv. Manuf. Tech. 1987; 2:75-88

Jiang W, Z Wang and Y Cai. Computer-aided group fixture design. Annals of the CIRP. 1988: 37(1);145-148

Imhof, G and W. Grahl. Die auswahl von vorrichtungselementen als bestandteil der rechnergestutzten projektierenden konstruktion von vorrichtungen, Wiss. Z. d. Techn. Hochsc. Karl-Marx-Stadt. 1977; 19(1):49-61

Ingrand, F and J Latombe. Functional reasoning for automatic fixture design. Lifia (IMAG) France, 1980.

Lim, BS and JAG Knight. Holdex - Holding device expert system. Proc. 1st Int. Conf. on Applications of Artificial Intelligence in Engineering Problems 1986:483-493

Markus, A, Z Markusz, J Farkas and J Filemon. Fixture design using PROLOG: An expert system. Robotics and Computer Integrated Manufacturing. 1984; 1(2):167-172.

Miller, AS and RG Hannam, Computer aided design using a knowledge base approach and its application to the design of jigs and fixtures. Proc. I.Mech.E. 1985; 199(B4)

Nee, AYC, N Bhattacharya, and AN Poo. Applying AI in jigs and fixtures design. Robotics and Computer Integrated Manufacturing. 1987; 3(2):195-201

Nee, AYC and AN Poo. Expert CAD systems for jigs and fixtures, Artificial Intelligence in Design (edited by D.T. Pham), IFS/Springer. 1990: 343-370

Nnaji BO, S Alladin and P Lyu. A framework for a rule-based expert fixturing system for face milling planar surfaces on a CAD system using flexible fixtures. J of Manu Sys. 1988: 7(3);193-207

Nnaji, BO and P Lyu. Rules for an expert fixturing system on a CAD screen using flexible fixture. J of Intelligent Manufacturing. 1990; 1:31-48.

Pham, DT and A De Sam Lazaro. Autofix - An expert CAD system for jigs and fixtures. Int. J. Mach. Tools Manufact. 1990;30(3):403-411.

Shin, H. Rule-based systems approach to fixture design. PhD Dissertation, Dept. of Industrial and Systems Engineering, The Ohio State University, Ohio, USA. 1988.

Trappey JC and CR Liu. A literature survey of fixture design automation. Int. J of Adv. Manuf. Technology. 1990:5(3):240-255

Young RIM and R Bell. Fixturing strategies and geometric queries in set-up planning. Int. J of Prod. Res. 1991: 29(3);537-550

Feature recognition and feature-based design for product model

Aldefeld, B. On automatic recognition of 3D structures from 2D representations. Computer Aided Design. 1983;15(2):59-64

Arbab, F. Requirements and architecture of CAM oriented CAD systems for the design and manufacture of mechanical parts. Ph.D thesis, Univ. of California, Los Angeles, 1982.

Braid, IC and CA Lang. Computer aided design of mechanical components with volume building bricks. Computer Languages for Numerical Control J., 1973:173-184

Choi, BK, MM Barash, and DC Anderson. Automatic recognition of machined surfaces from a 3D solid model. CAD Journal. 1984;16(2):81- 86

Dong. X, DeVries, WR and WJ Wozny. Feature-based reasoning in fixture design. Annals of the CIRP. 1991;40(1):111-114.

Fu, KS. Syntactic Pattern Recognition and Applications, Englewood Cliffs, N.J., Printice-Hall, 1982.

Grayer, AR. The automatic production of machined components staring from stored geometric description., in Mcpherson, D(ed) Advances in Computer-Aided Manufacturing, North-Holland Publishing Co., 1977:137-151

Haralick, RM and D Queeney. Understanding engineering drawings. Computer Graphics and Image processing, 1982;20: 244-258

Henderson, MR and DV Anderson. Computer recognition and extraction of form features: A CAD/CAM link. Computers in Industry 1984;5(4):329- 339

Henderson, MR. Extraction of feature information from three dimensional CAD data. Ph.D Dissertation, Purdue University, USA, May, 1984.

Hirschtick, JK and DC Gossard. Geometric reasoning for design advisory systems. Proc. of the ASME Int. Computers in Engineering Conf. July 20-24, 1986:1:263-270

Ingrand, F and J Latombe. Functional reasoning for automatic fixture design. Lifia (IMAG) France, 1980.

Jakubowski, R. Syntactic characterization of machine parts shapes. Cybernetics and Systems Int. J. 1982;13:1-24

Jared, GEM. Introduction to solid modelling. in Earnshaw, R.A., Parslow, R.D. and Woodark, J.R. (eds) Geometric Modelling and Computer Graphics-Techniques and Applications, Technical press-Unicom, UK, 1987.

Jared, GEM. Shape features in geometric modelling, in Pickett, M.S. and Boyes, J.W. (eds) Solid Modelling by Computers: From Theory to Application, Plenum, USA, 1984.

Joshi, S and TC Chang. Graph-based heuristics for recognition of machined features from a 3D solid model. Computer-Aided Design, March 1988:20(2):58-66

Joshi, S. CAD interface for automated process planning. Ph.D Dissertation, Purdue University, USA, 1987.

Kung, H. An investigation into development of process plans from solid geometric modelling representation. Ph.D Dissertation, Oklahoma State University, USA, 1984.

Lee YC and KS Fu. Machine understanding of CSG: extraction and unification of manufacturing features. IEEE Computer Graphics and Applications.1987; 7(1):20-32

Luby SC, JR Dixon and MK Simmons. Creating and using a features database. Computers in Mechanical Engineering. 1986:5(3)

Mantyla, M and R Sulonen. GWB: A solid modeller with Euler operators. IEEE Computers graphics and applications. Sept. 1982:2(7): 17-31

Markowsky, G and MA Wesley. Fleshing out wire frames. IBM J. of Research and Development. 1980:24(5):582-597

Okino N, Y Kakazu and Kubo. TIPS-1: Technical information processing system for computer aided design, drawing and manufacturing. Computer Languages for Numerical Control J. 1973:141-150

Senthil kumar A. A knowledge-based generative modular fixture design methodology for prismatic workpieces. Ph.D thesis, National University of Singapore, Singapore. 1992

Srinivasan, R, CR Liu, and KS Fu. Extraction of manufacturing details from geometric models. Comput. & Indust. Engg. 1985:9(2):125-133

Staley, S, M Henderson and D Anderson. Using syntactic pattern recognition to extract feature information form a solid geometry model data base. Computers in Mechanical Engineering. Sept. 1983: 61-66

Woo, TC. Computer aided recognition of volumetric designs. in Mcpherson, D (ed) Advances in computer-aided manufacturing, North-Holland Publishing Co., 1977:121-135

Woodwark, JR. Some speculations on feature recognition. Computer-Aided Design. 1988:20(4):189-196

Modular, phase-change and reconfigurable fixtures

Asada, H and A Fields. Design of flexible fixtures reconfigured by robot manipulators, Proceedings of ASME Winter Annual Meeting, Sym. on Robotics and Automation, 1985; 251-257.

Benhabib, B, KC Chan and MQ Dai. A modular programmable fixturing system. J of Engineering for Industry. 1991; 113:93-100.

Chan, KC, B Benhabib and MQ Dai. A reconfigurable fixturing system for robotic assembly. J. of Manuf. Sys., 1990;9(3):206-221.

Gandhi, MV and BC Thompson. Phase-change fixturing for flexible manufacturing systems. J of Manu. Sys. 1985; 4(1):29-39.

Giusti, F., M. Santochi, and G. Dini, "Robotised assembly of modular fixtures", Annals of the CIRP, Vol.40, No.1, pp.17-20, 1991.

Hoffman EG. Modular fixturing. Manufacturing technology Press, Lake Geneva, Wisconsin, 1987

Horie T. Adaptability of a modular fixturing system to factory automation. Bull. Japan Soc. of Prec. Engg. 1988:22(1);1-5

Jeries AH, K Okamura and T McGreevy. Dynamic behaviour and creep characteristics of flexible particulate bed fixtures. J of Manu Sys. 1993: 12(6);496-505

Thompson BS, MV Gandhi and DJ Desai. Workpiece-fixture interactions in a compacted fluidized fixture under various loading conditions. Int. J of Prod. Res. 1989: 27(2); 229-246

Thompson, B.C. and M.V. Gandhi, "Commentary on flexible fixturing", ASME Appl. Mech. Rev., Vol.39, No.9, pp 1365-1369, Sept. 1986.

Xu, Y., " A modular fixturing system (MFS) for flexible manufacturing", The FMS Magazine, pp.292-296, Oct. 1983.

Youcef-Toumi, K, JJ Bausch and SJ Blacker. Automated setup and reconfiguration for modular fixturing. Proceedings of Manufacturing Int. '88, Symp. on Manuf. Sys. - Design, Integration, and Control, ASME, April 17-20. 1988;III:121-128

Youcef-Toumi, K, WS Liu and H Asada. Computer-aided analysis of reconfigurable fixtures and sheet metal parts for robotic drilling. Proc. of the Int. Conf. on the Manufacturing Science and Technology of the Future, 1987.

Youcef-Toumi, K. and J.H. Buitrago, "Design of robot-operated adaptable fixtures", Proceedings of Manufacturing Int. '88, Symp. on Manuf. Sys. - Design, Integration, and Control, Vol. III, ASME, April 17-20, pp.113-119. 1988.

Tolerance control

ANSI Y14.5-1982, "Dimensioning and tolerancing", ASME, NY, USA, 1982.

Bennett B and LO Gupta. Least-cost tolerance-Part I, II. Int. J. of Prod. Res., 1970;8: 65-71

Britton GA, Whybrew K, Sermsuti-anuwat Y. A manual graph-theoretic method for teaching tolerance charting. International Journal of Mechanical Engineering Education October 1992; 20(4):273-285.

Eary DF, Johnson GE. Process engineering for manufacturing. Englewood Cliffs, NJ, Prentice-Hall, 1962.

Gadzala JL. Dimensional control in precision manufacturing. McGraw-Hill, 1959.

Ling KC. Development of computerized tolerance charting system. Project report, School of Mechanical and Production Engineering, Nanyang Technological University, 1992.

Marks CT. Tolerance charts control production machining. American Machinist March 1953; 97(5) :114-116

Nee, AYC, N Bhattacharya and AN Poo. Computer-aided datum determination using the least tolerance method. SME Technical Report, MS87- 905, 1987.

Ngoi NK, Whybrew K. A fast spatial representation method (applied to fixture design), International Journal of Advanced Manufacturing Technology 1993; 8(2):71-77

Robinson DF, Foulds LR. Digraph: theory and techniques. Gordon and Breach Science Publishers, New York, 1980.

Roy U and CR Liu. Feature-based representational scheme of a solid modeller for providing dimensioning and tolerancing information. Robotics and Computer-Integrated Manufacturing. 1988;4(3/4):335-345

Senthil kumar A. and AYC Nee. Tolerance analysis in modular fixture design. Journal of the Inst. of Engrs. Singapore. 1990;30(1):40-50

Wade OR. Tolerance control in design and manufacturing. Industrial Press, New York, 1967.

Whybrew K, Britton GA, Robinson DF, Sermsuti-anuwat Y. A graph-theoretic approach to tolerance charting. International Journal of Advanced Manufacturing Technology. May 1990; 5(2):175-183

Fixture design analysis

Chou, Y-C, V Chandru and MM Barash. A mathematical approach to automatic configuration of machining fixtures: analysis and synthesis. Trans. of the ASME Journal of Engineering for Industry, November 1989;111:299-306

Daimon, M., T. Yoshida, N. Kojima, H. Yamamoto and T. Hoshi, Study for designing fixtures considering dynamics of thin-walled plate- and box-like workpieces, Annals of the CIRP Vol.34/1/1985, pp.319-322.

Fuh JY, CH Chang, MA Melkanoff and H Rau. CAFP: Computer-aided fixture planning. Proc of the computers in engineering, ASME, 1993:1-10

King LSB and I Hutter. Theoretical approach for generating optimal fixturing locations for prismatic workparts in automated assembly. J of Manu Sys. 1993: 12(5);409-416

King, LSB., M Hart and D Wong, Geometry and physics of automatic fixturing for assembly, Proceedings of the Int Conf on CIM (ICCIM'91), Singapore. 2-4 October 1991; 525-529

Lee, JD and LS Haynes. Finite-element analysis of flexible fixturing system. Trans. of the ASME Journal of Engineering for Industry, May 1987; 109:134-139

Lee, SH and M Cutkosky. Fixture planning with friction. 1989 NSF Design Theory and Methodology Conference, Amherst, MA. June, 1989:1-28

Menassa RJ and WR DeVries. Optimization methods applied to selecting support positions in fixture design. Tran. of the ASME, J of Engg for Ind. 1991:113;412-418

Menassa, RJ and WR DeVries. Locating point synthesis in fixture design. Annals of the CIRP, 1989;38(1):165-169

Menassa, RJ and WR DeVries. Optimization methods applied to selecting support positions in fixture design. USA-Japan Symposium on Flexible Automation, 1988;1:457-482

Mittal, RO, PH Cohen and BJ Gilmore. Dynamic modeling of the fixture-workpiece system. Robotics & Computer-Integrated Manufacturing. 1991;8(4): 201-217

Index

MIX
Papier aus verantwortungsvollen Quellen
Paper from responsible sources
FSC® C105338

If you have any concerns about our products,
you can contact us on
ProductSafety@springernature.com

In case Publisher is established outside the EU,
the EU authorized representative is:
Springer Nature Customer Service Center GmbH
Europaplatz 3, 69115 Heidelberg, Germany

Printed by Libri Plureos GmbH
in Hamburg, Germany